Lecture Notes in Mathematics

Edited by A. Dold and B. Eckmann

557

Wolf Iberkleid
Ted Petrie

Smooth S^1 Manifolds

Springer-Verlag
Berlin · Heidelberg · New York 1976

Authors
Wolf Iberkleid
CENTRO de INV. del IPN
Apdo. Postal 14740
Mexico, 14 DF/Mexico

Ted Petrie
Rutgers University
Department of Mathematics
New Brunswick, N. J./USA

Library of Congress Cataloging in Publication Data

Iberkleid, Wolf, 1946-
Smooth S^1 manifolds.

(Lecture notes in mathematics ; 557)
Bibliography: p.
Includes index.
1. Differential topology. 2. Manifolds (Mathematics) 3. Characteristic classes. 4. Topological transformation groups. I. Petrie, Ted, 1939- joint author. II. Title. III. Series: Lecture notes in mathematics (Berlin) ; 557.
QA3.L28 no. 557 [QA613.6] 510'.8s [514'.7] 76-50065

AMS Subject Classifications (1970): 57D20, 57D65, 57E25, 55B25

ISBN 3-540-08002-3 Springer-Verlag Berlin · Heidelberg · New York
ISBN 0-387-08002-3 Springer-Verlag New York · Heidelberg · Berlin

Printed in Germany.
Printing and binding: Beltz Offsetdruck, Hemsbach/Bergstr.

Smooth S^1 Manifolds

by Wolf Iberkleid and Ted Petrie

INTRODUCTION

Part I

We single out a set $\underline{P}$ of prime ideals in $R(S^1)$ (the complex representation ring of S^1) and consider an $R(S^1)_P$ (localization at P) valued bilinear form on $K^*_{S^1}(X)_P$ the localized complex equivariant K theory of the smooth closed S^1 manifold X. Our primary interest is to show that when the underlying smooth manifold to X, written $|X|$, is a spin^c manifold, the bilinear form is non-degenerate and to study the geometric consequences of this fact. Of secondary interest at <u>this time</u> is the analogous study for other compact Lie groups G. When we can offer insight on smooth G manifolds without disrupting our main discussion of $G = S^1$ we do so.

Here in more detail is the definition of the bilinear form. The ring $R(S^1) = Z[t,t^{-1}]$ contains the integers Z. Let P_Z denote the set of prime ideals of $R(S^1)$ generated by a prime of Z. In the subsequent discussion P denotes any set of prime ideals of $R(S^1)$ such that $P \cap P_Z = \phi$. The set of all prime ideals of $R(S^1)$ minus P_Z is denoted by P_O. Let $R = R(S^1)_P$ denote the localization of $R(S^1)$ at the set of primes P. Specifically if $\bar{F}$ denotes the field of fractions of $R(S^1)$, then $R \subset \bar{F}$ consists of "fractions" a/b with b prime to all the ideals in P. In particular for $P = P_O$, $R = R(S^1) \underset{Z}{\otimes} Q$.

The complex equivariant K theory of the S^1 space X is denoted by $K^*_{S^1}(X) = K^0_{S^1}(X) \oplus K^1_{S^1}(X)$ [3]. It is a module over $R(S^1) = K^0_{S^1}(p)$ where p is a one point space. The localized equivariant K theory $K^*_{S^1}(X)_P$ is an R module as is $K^*_{S^1}(TX)_P$ where TX is the tangent space of X. (Note $K^*_{S^1}(X)_{P_0} = K^*_{S^1}(X) \otimes_Z Q$.) In fact $K^*_{S^1}(TX)_P$ is a module over $K^*_{S^1}(X)_P$ and there is a distinguished homomorphism

$$Id^X_R : K^0_{S^1}(TX)_P \longrightarrow R$$

induced by the Atiyah-Singer index homomorphism

$$Id^X_{S^1} : K^0_{S^1}(TX) \longrightarrow R(G) \quad [5].$$

Define

$$\Phi^X_R : K^*_{S^1}(TX)_P \longrightarrow \mathrm{Hom}_R(K^*_{S^1}(X)_P, R)$$

by

$$\Phi^X_R(x)[y] = Id^X_R(x \cdot y)$$

$x \in K^*_{S^1}(TX)$, $y \in K^*_{S^1}(X)$. Note that Φ^X_R preserves degree, i.e.,

$$\Phi^X_R : K^i_{S^1}(TX)_P \longrightarrow \mathrm{Hom}_R(K^i_{S^1}(X)_P, R);\ i = 0,1.$$

The most important step in constructing the desired bilinear form is to show

Theorem 5.6. There is a split exact sequence

$$0 \longrightarrow \mathrm{Ext}^1_R(h^{*+1}(X), R) \longrightarrow h^*(TX) \xrightarrow{\phi^X_R} \mathrm{Hom}_R(h^*(X), R) \longrightarrow 0$$

where $h^*(X) = K^*_{S^1}(X)_p$.

The easier step is in constructing an isomorphism

$$\Delta^X : K^*_{S^1}(X)_p \longrightarrow K^*_{S^1}(TX)_p$$

when $|X|$ is a spin^c manifold. The desired bilinear form on $K^*_{S^1}(X)_p$ written $\langle\ \rangle_X$ is defined by

$$\langle a,b\rangle_X = Id^X_R(\Delta^X(a)\cdot b).$$

From Theorem 5.6 it follows that the induced bilinear form on $K^*_{S^1}(X)_p$ modulo its R torsion subgroup is non-degenerate (Theorem 6.19).

Because of the analogy with the cup product bilinear form on the middle dimensional cohomology of an even dimensional manifold and because of the rich arithmetical structure of the ring $R(S^1)$, one might expect that the bilinear form on $K^*_{S^1}(X)_p$ would have some powerful geometric consecuences. We indicate some of the geometrical consequences in the second part of these notes. **For an outline of these see end of Introduction to Part II.**

Here are the ideas involved in proving Theorem 5.6 . The proof is divided into two parts. This dichotomy is due to the difference between orbits of points $x \in X^{S^1}$ and orbits of points while in the second orbits are circles. Homologically, this geometric fact is translated into

(i) $K^{*}_{S^1}(O_x)_p$ is a free R module if $x \in X^{S^1}$

(ii) $K^{*}_{S^1}(O_x)_p$ is a torsion R module if $x \in X - X^{S^1}$.

Here O_x is the orbit of the point x.

Once this difference in character between X^{S^1} and $X - X^{S^1}$ is realized, it is convenient to introduce a non trivial "index" for fixed point free manifolds

$$Id^{X}_{F/R} : K^{1}_{S^1}(TX)_p \longrightarrow F/R$$

where F is a ring satisfying $R \subset F \subset \bar{F}$. This induces a homomorphism

$$\phi^{X}_{F/R} : K^{*}_{S^1}(TX)_p \longrightarrow \mathrm{Hom}_R(K^{*+1}_{S^1}(X)_p, F/R).$$

We are now led into a more general situation. Let us introduce the notation

$$h^{*}(X) = K^{*}_{S^1}(X)_p$$

Homology is only defined for compact smooth S^1 manifolds and is defined in such a way that if X is a closed S^1 manifold,

$$h_{*}(X) = h^{*}(TX)$$

and cap product is defined by the structure of $h^*(TX)$ as a module over $h^*(X)$. In terms of this notation we get homomorphisms

$$\phi_C^{(X,Y)} : h_*(X,Y) \longrightarrow \mathrm{Hom}_R(h^*(X,Y),C)$$

for $C = R, F/R$, given by

$$\phi_C^{(X,Y)}(a)[b] = \mathrm{Id}_C(a \cap b).$$

Note that $\phi_{F/R}^X$ is defined only when $X^{S^1} = \emptyset$.

We show that there is a map of short exact sequences

$$\begin{array}{ccccccccc}
0 & \longrightarrow & h_*(X^{S^1}) & \xrightarrow{i_*} & h_*(X) & \xrightarrow{j_*} & h_*(X,X^{S^1}) & \longrightarrow & 0 \\
 & & \Big\downarrow \phi_R^{X^{S^1}} & & \Big\downarrow \phi_R^X & & \Big\downarrow \Delta & & \\
 & & & & & & \mathrm{Hom}_R(h^{*+1}(X,X^{S^1}),F/R) & & \\
 & & & & & & \Big\downarrow i^* & & \\
0 & \longrightarrow & \mathrm{Hom}_R(h^*(X^{S^1}),R) & \longrightarrow & \mathrm{Hom}_R(h^*(X),R) & \xrightarrow{\omega} & \mathrm{Hom}_R(K,F/R) & \longrightarrow & 0
\end{array}$$

where Δ is $\phi_{F/R}^{(X-\dot{N},\partial N)}$ after we identify $h(X,X^{S^1}) = h(X,N) = h(X-\dot{N},\partial N)$. N denotes a closed tubular neighborhood of X^{S^1}, and $K \subset h^{*+1}(X,X^{S^1})$ is the image of the coboundary homomorphism.

We show that $\phi_R^{X^{S^1}}$ is an isomorphism by a simple computation using the formula of Atiyah-Singer for the homomorphism $\mathrm{Id}_R^{X^{S^1}}$. We prove that $\phi_{F/R}^{(X-\dot{N},\partial N)}$ is an isomorphism by using the theorem of Wasserman which asserts that X is obtained from N by attaching S^1 handles over orbits of points of $X-\dot{N}$. An S^1 handle is essentially the disk bundle of a real S^1 vector bundle over an orbit. These are easy to analyze geometrically and algebraically. Using this analysis, we prove that $\phi_{F/R}^{(X-\dot{N},\partial N)}$ is an isomorphism by induction on the handles.

1. Preliminary remarks on G spaces.

We discuss the general setting of G spaces. In particular we mention some properties of a cohomology theory $h^*(\ )$ on the category of locally compact G spaces. The theory we have in mind is either $h^* = K_G^*$ or obtained from K_G^* in some straight forward way. Specifically for $G = S^1$, $h^* = K_{S^1}^*(\ )_P$. The <u>essential idea</u> of this section is <u>to define a homology theory</u> $h_*(\ )$ on the subcategory of compact smooth G manifolds which is dual to $h^*(\ )$. This is done by means of the tangent space of a smooth G manifold. If X is a smooth G manifold and p is a point with trivial G action, there is a unique G map $\rho^X: X \to p$ and the induced homomorphism $\rho_0^X: h_0(X) \to h_0(p) = R$ is called the index homomorphism and denoted by Id_R^X.

2. Structure of smooth G manifolds in terms of handles.

The basic structure theorem about smooth G manifolds is that such a manifold is a union of G handles. When $G = S^1$

we give a very precise geometric description of S^1 handles and an algebraic description (in terms of cohomology) of the S^1 orbit associated to the handle.

3. Multiplicative properties of $h^*(\ )$.

We discuss cross and cap products and their functional properties.

4. Fixed point free actions.

We introduce the torsion index

$$Id_{F/R}:h_1(X) \to F/R$$

defined on fixed point free actions. Cap product and $Id_{F/R}$ induce a homomorphism

$$\phi_{F/R}^{(X,\partial X)}:h_*(X,\partial X) \longrightarrow Hom_R(h^{*+1}(X,\partial X),F/R)$$

which we show is an isomorphism. To prove this we study some homological properties of F/R and then use a handle decomposition of X and induction to get the result. With a handle decomposition of X the problem is reduced to proving the isomorphism

$$\omega_\Gamma^V:K_\Gamma^*(V) \longrightarrow Hom_{R(\Gamma)}(K_\Gamma^*(V),R(\Gamma))$$

which is induced from cross product and the Thom isomorphism. Here Γ is a discrete subgroup of S^1 and V a real Γ module.

The fact that $\phi_{F/R}^{(X,\partial X)}$ is an isomorphism is the first of two key steps in showing the universal coeficient theorem.

5. The universal coefficient theorem.

The second key step in establishing the u.c.th. is to show that

$$\phi_R^{X^{S^1}}:h_*(X^{S^1}) \longrightarrow \mathrm{Hom}_R(h^*(X^{S^1}),R)$$

is an isomorphism. This we do by noting that the problem can be translated into a non equivariant one. We then show that diagram 0.1 is commutative. This tells us that ϕ_R^X is an epimorphism. The left part of the u.c.th. is a trivial consequence of homological algebra.

6. Poincare duality.

The final step showing that the $R = R(S^1)_p$ valued bilinear form on $h^*(X) = K^*_{S^1}(X)_p$ is non-degenerate when X is a smooth closed S^1 manifold is to establish a Poincare duality isomorphism

$$\Delta^X:h^*(X) \longrightarrow h_*(X).$$

Then the bilinear form on $h^*(X)$ is defined by

$$\langle a,b\rangle_X = \Phi_R^X(\Delta^X(a))[b]$$

and is non-degenerate.

The main theorem can be analyzed for other compact connected Lie groups G by replacing S^1 by G, P by some set of prime ideals P' in $R(G)$ and R by $R(G)_{P'}$. We suspect that the question of surjectivity in this case depends upon the homological dimension of $K_G^*(X)_{P'}$ as an $R(G)_{P'}$ module. For more complicated groups G (where the global dimension of $R(G)_{P'}$ exceeds 1) there should be a spectral sequence relating h_*^G and h_G^* where $h_G^* = K_G^*(\ )_{P'}$. In this more general setting our theorem will become a statement that the spectral sequence collapses.

At the end of Section 5 we indicate how the u.c.th. can be generalized to the n dimensional torus T^n if the localization $R(T^n)_P$ has global dimension 1.

For additional information concerning analogs of the bilinear form $\langle\ \rangle_X$ on $K_{S^1}^*(X)_P$ one should consult [9] and [13].

Concerning the style of the paper. Much of the material of this paper is involved with general properties of a cohomology theory on a category. These do not always depend upon the fact that the cohomology theory is K_G^* or some simple variant. In order to facilitate notation we denote the cohomology theory by

h^* and list the assumptions concerning the theory we need. We emphasize that our applications deal with $h^* = K^*_{S^1}(\ )_P$ or $K^*_{\Gamma}(\ )_P$ $\Gamma \subset S^1$ and in order not to become involved too much in axiomatics we assume properties for this theory which hold for equivariant K theory.

Part II

The material of Part I provides the tools for the study of smooth actions of a compact Lie Group G. The main questions to which we address ourselves are these: Let M and N be two homotopy equivalent manifolds. (1) Suppose there is a G manifold Y with $|Y| = M$. Is there a G manifold X with $|X| = N$? (2) Given Y with $|Y| = M$ how can we construct X with $|X| = N$? The central question which must be answered for dealing with (1) and (2) is: What are the relations among the representations of G on the tangent spaces TY_p, $p \in Y^G$ and the global invariants of $|Y|$ e.g. its Pontrjagin classes and cohomology?

Example 1. Global assumption: X is a smooth closed manifold with $H^*(X,Q) = H^*(S^{2n},Q)$. Suppose that our compact group G acts on X with just 2 fixed points p and q and assume that the action is free outside p and q.

Conclusion: Atiyah-Bott [3]; the two real representations of G on the tangent space to X at p and q are equal. Thus a cohomological assumption implies an equality of representations at the tangent spaces at the fixed points.

Example 2. Global Assumption: X is a closed manifold having the same cohomology ring as complex projective n space.

Suppose S^1 acts on X and the fixed point set consists of isolated points. Then the collection of representations of S^1 on the tangent space at the various fixed points determine all the Pontrjagin classes of X [9].

For dealing with these questions, we introduce the set $S_G(Y)$ attached to the closed G manifold Y. It consists of equivalence classes of pairs (X,f) where X is a closed G manifold and $F : X \dashrightarrow Y$ is a map such that

(i) $|f| : |X| \dashrightarrow |Y|$ is a homotopy equivalence.

Here $|f|$ means the underlying map to f obtained by neglecting its relation to G.

(ii) $|f^G| : |X^G| \dashrightarrow |Y^G|$ is a homotopy equivalence. Two pairs (X_i, f_i) $i = 0,1$ are equivalent if there is a G homotopy equivalence $\phi : X_0 \dashrightarrow X_1$ such that $f_1 \phi$ is G homotopic to f_0. The equivalence class of (X,f) is denoted by $[X,f] \in S_G(Y)$. The identity map of Y is denoted by I_Y and the element $[Y,I_Y] \in S_G(Y)$ is called the trivial element.

Condition (i) imposes stringent restrictions on $\ker f^*$ and $\operatorname{coker} f^*$ where f^* is the induced map in equivariant K theory. The essential algebraic fact here is the Localization Completion Lemma 3.2. When $G = S^1$, we connect the algebras $K^*_{S^1}(X)$ and $K^*_{S^1}(Y)$ with the S^1 representations $\left\{TX_p | p \in X^{S^1}\right\}$ and $\left\{TY_q | q \in Y^{S^1}\right\}$ and with the Pontrjagin classes of $|X|$ and $|Y|$. The invariant of $[X,f] \in S_{S^1}(Y)$ which gives these connections is the torsion (see 2.4)

$$f_*(1_X) \in \widetilde{K}^*_{S^1}(Y).$$

If the torsion is the identity of the algebra $\widetilde{K}^*_{S^1}(Y)$, then $TX_p = TY_{f(p)}$ $p \in X^{S^1}$ and $|f|^*$ preserves Pontryjagin classes provided $|Y|$ satisfies suitable hypotheses. In particular if f is an S^1 homotopy equivalence, $|f|^*$ preserves Pontryjagin classes (Theorem 6.4).

Here is a useful example:

Let M and N denote the following complex S^1 modules of dimension 2.

(i) $|M| = C^2$ and if $(z_1, z_2) \in M$ denote the complex coordinates of a point $z \in M$ and $t \in S^1$, $t(z_1,z_2) = (tz_1, t^{pq}z_2)$ where p,q are positive relatively prime integers and tz_1, $t^{pq}z_2$ denote the indicated products using complex multiplication.

(ii) $|N| = C^2$ and for $z = (z_1,z_2) \in N$ $t(z_1,z_2) = (t^p z_1, t^q z_2)$.

Choose positive integers a and b which satisfy $-ap+bq = 1$ and define an S^1 map $\omega : N \longrightarrow M$ by

$$\omega(z_1,z_2) = (\bar{z}_1^a z_2^b, z_1^q + z_2^p)$$

where $\bar{z}_1$ denotes the complex conjugate of z_1. Observe that ω is proper, so there is an induced S^1 map $\omega^+ : N^+ \longrightarrow M^+$

between the one point compactications M^+ and N^+ of M and N. Observe also that $|M^+| = |N^+| = S^4$ (the four sphere) and that the S^1 actions defined by M^+ and N^+ are smooth. One checks that $|\omega^+|$ is a degree one map and clearly

$\omega^{+S^1} : (N^+)^{S^1} \longrightarrow (M^+)^{S^1}$ is a homotopy equivalence; so

$$[N^+, \omega^+] \in S_{S^1}(M^+).$$

I claim $[N^+, \omega^+] \neq [M^+, I_{M^+}]$. This fact is left to the reader. We shall encounter this example again in the development of more sophisticated ideas.

Our setting for studying G actions on manifolds homotopy equivalent to a fixed manifold M involves:

(a) Find a complete set of universal G models for M. This means to find a collection of S^1 manifolds X_α such that $|X_\alpha|$ is homotopy equivalent to M and if X is an S^1 manifold for which $|X|$ is homotopy equivalent to M, then for some map f and some α, $[X,f] \in S_G(X_\alpha)$.

(b) Determine the properties of the set $S_G(X_\alpha)$. Specifically compare the local properties and global properties of X and X_α when $[X,f] \in S_G(X_\alpha)$. E.g., compare TX_p and $T(X_\alpha)_{f(p)}$ for $p \in X^{S^1}$ and compare the Pontryagin classes of $|X|$ and $|X_\alpha|$.

As an application of the preceding ideas, we study S^1 actions on $P(C^n)$, the space of complex lines in C^n. In particular for certain S^1 modules Ω we produce a non trivial element $[X(\omega),\bar{\omega}] \in S_{S^1}(P(\Omega))$ and illustrate the necessity of the hypothesis in the theorems comparing the local and global properties of X and Y when $[X,f] \in S_{S^1}(Y)$. The S^1 manifolds $X(\omega)$ also show the necessity of the hypothesis of Bredon [8] concerning the structure of fixed points sets of subgroups of S^1.

Let me mention another interesting feature of the manifolds $X(\omega)$. Observe that if $[X,f] \in S_G(Y)$ and if E is an acyclic space on which G acts freely, then

$$E \times_G X \text{ and } E \times_G Y$$

are homotopy equivalent. Thus any homotopy functor such as $H^*(\ )$ or $K^*(\ )$ applied to these spaces is inadequate to distinguish the G manifolds X and Y, in particular, it cannot distinguish the representations of S^1 on the tangent spaces TX_p and $TY_{f(p)}$ for $p \in X^{S^1}$. Specifically, for the element $[X(\omega),\bar{\omega}] \in S_{S^1}(P(\Omega))$, we have $TX(\omega)_p \neq TP(\Omega)_{f(p)}$ for all $p \in X(\Omega)^{S^1}$. In fact there is no linear action of S^1 on complex projective space $P(C^{n+1})$ $(n+1 = \dim \Omega)$ whose collection of representations of S^1 on the tangent spaces at the isolated fixed points agree with the collection $\{TX(\omega)_p | p \in X^{S^1}\}$. In particular, the <u>ring</u> $H^*(E \times_{S^1} X(\omega))$ does not determine these representations as suggested by some authors. On-the-other-hand, the torsion $\tilde{\omega}_*(1_{X(\omega)}) \in K^*_{S^1}(P(\Omega))$ detects the difference between these representations in a strong way. See (ii) and (iii) above.

The first eight sections of Part II deal with the material discussed above. Section 9 presents some applications of the ideas of Part I section 4 to the study of S^1 actions with finite isotropy groups. Section 10 abstracts the algebraic situation which arises from the bilinear form on $K^*_{S^1}(X)$. As an interesting applications of the interplay between the abstract algebraic situation and the geometry see Proposition 10.14 and Remark 10.15.

Consequences of non-singularity (Outline)

The facts that $< \ >_X$ and $< \ >_Y$ are non-singular for S^1 manifolds X and Y implies that any S^1 map $f: X \to Y$ gives rise to a homomorphism of R modules $f_*: h^*(X)/T_X \to h^*(Y)/T_Y$ where T_X resp. T_Y is the R torsion submodule of $h^*(X)$ resp. $h^*(Y)$. (II 2.4). It is defined by

$$<f_*(x),y>_Y = <x,f^*(y)>_X \quad \text{for}$$

$x \in h^*(X)/T_X$ and $y \in h^*(Y)/T_Y$ and satisfies

(i) $$f^*f_*(x) = f^*f_*(1_X)\cdot x$$

where 1_X is the identity of $h^*(X)/T_X$. In-other-words non-singularity of $< \ >_X$ and $< \ >_Y$ gives an isomorphism

(ii) $$h^*(X)/T_X \cong \mathrm{Hom}_R(h^*(X)/T_X, R) \quad \text{and}$$

similarly for Y. Using these isomorphisms and their inverses gives

(iii) $$f_* = \mathrm{Hom}_R(f^*,R)$$

In particular (i) and (ii) imply (iii) If f is an isomorphism, then

$$f^* \quad \text{is an isomorphism and} \quad f_*(1_X)$$

is a unit of $h^*(Y)/T_Y$.

Suppose X and Y satisfy the hypothesis (H) of II §2 and $f: X \to Y$ is an S^1 map with $|f|$ a homotopy equivalence and that X^{S^1} is a isolated set of points. Since $f^{S^1}: X^{S^1} \to Y^{S^1}$ is a homotopy equivalence, it is a homeomorphism. There is a very important relationship between: The S^1 modules $\{TX_p, TY_{f(p)}, p \in X^{S^1}\}$, the algebras $h^*(X)/T_X$, $h^*(Y)/T_Y$ and the cohomology classes $\hat{A}(|Y|)$, $\hat{A}(|X|)$ (II 4.3) which determine the Pontrjagin classes of $|X|$ and $|Y|$. This relationship is exhibit by (iv)-(vi):

(iv)

$$\begin{array}{ccccc} 0 & & & & \\ \downarrow & & & & \\ h^*(Y)/T_Y & \xrightarrow{\epsilon} & K^*(Y) \otimes Q & \xrightarrow{ch} & H^*(Y,Q) \\ \downarrow i^* & & & & \\ h^*(Y^{S^1}) & & & & \end{array}$$

(v)
$$ch\,\epsilon\, f_*(1_X) = g^*\hat{A}(|X|)/\hat{A}(|Y|)$$

$g: Y \to X$ is a homotopy inverse of f.

(vi)
$$i^*_{f(p)} f_*(1_X) = (\mathrm{unit}_{f(p)}) \cdot \lambda_{-1}(T\hat{Y}_{f(p)})/\lambda_{-1}(T\hat{X}_p)$$

for each $p \in X^{S^1} \cong Y^{S^1}$. Here $\mathrm{unit}_{f(p)} \in R$ is a unit, i^*_q denotes the composition $h^*(Y)/T_Y \xrightarrow{i^*} h^*(Y^{S^1}) \to h^*(q)$ for $q \in Y^{S^1}$ and $T\hat{X}_p$ is a complex S^1 module whose underlying real S^1 module is TX_p.

Since $h^*(Y^{S^1}) \cong \prod_{q \in Y^{S^1}} h^*(q)$ and since i^* is a monomorphism it follows from (vi) that $f_*(1_X)$ is determined by $\{T\hat{Y}_{f(p)}, T\hat{X}_p \mid p \in X^{S^1}\}$ modulo the units $\mathrm{unit}_{f(p)}$ and conversely; moreover, there is enough control over the units $\mathrm{unit}_{f(p)}$ $p \in X^{S^1}$ to assert that $ch\,\epsilon\, f_*(1_X)$ is determined by the representations $\{TX_p, TY_{f(p)} \mid p \in X^{S^1}\}$ i.e. $g^*A(|X|)/A(|Y|)$ is determined by this data because of (iv)-(vi).

The above relationship between the algebras, the representatio and the $\hat{A}$ classes allows deductions like these (II 6.4): If If $f^*: h^*(Y)/T_Y \to h^*(X)/T_X$ is an isomorphism (e.g. if f is an

S^1 homotopy equivalence), then f_* is an isomorphism and $f_*(1_X)$ is a unit of $h^*(Y)/T_Y$ (iii). Thus $i^*_{f(p)}\, f_*(1_X)$ is a unit of R and by (vi) this implies $TY_{f(p)} = TX_p$ for all $p \in X^{S^1}$; so by the above discussion, $ch \in f^*(1_X) = 1$ i.e. $g^*\hat{A}(|X|) = \hat{A}(|Y|)$ or g^* preserves Pontrjagin classes.

Each cyclic subgroup $Z_m \subset S^1$ determines a prime ideal $p_m \in R$ namely those characters which vanish on Z_m. If we know that f^* induces an isomorphism when $h^*(Y)/T_Y$ is localized at p_m (always true when m is a prime power), then $f_*(1_X)$ is a unit in this localized algebra and from (vi) we see that $\lambda_{-1}(T\hat{Y}_{f(p)})/\lambda_{-1}(T\hat{X}_p)$ is a unit of R_{p_m} for all $p \in X^{S^1}$. This implies that $\dim\ T\hat{X}_p^{Z_m} = \dim\ T\hat{Y}_{f(p)}^{Z_m}$ for each $p \in X^{S^1}$.

Here is another relation between the algebra and geometry which is a consequence of the non singularity of the bilinear form. Let ord_R (coker f^*) denote the ideal in R which annihilates coker f^*; $f^*: K^*(Y)/T_Y \to K^*(X)/T_X$ where $f: X \to Y$ is an S^1 map for which f^* induces an isomorphism over the field of fractions of R. E.g. X is Y^{S^1} and f the inclusion. Let $\det f_*(1_X)$ denote the determinant of the R endomorphism of $K^*(Y)/T_Y$ defined by multiplication by $f_*(1_X)$. Then the principle ideal defined by $\det f_*(1_X)$ and $[\mathrm{ord}_R \text{ coker } f^*]^2$ are equal (II 10.5). Suppose now Y^{S^1} consists of isolated points. For each $p \in Y^{S^1}$ set

$$T\hat{Y}_p = \sum_{i=1}^{d/2} t^{\lambda_i(p)} \qquad d = \dim Y$$

Then the absolute value of each integer $|\lambda_i(p)|$ occurs an even number of times in the collection $\{|\lambda_i(p)|\ |\ p \in Y^{S^1},\ i = 1,2,\ldots d/2\}$ (II 10.12). This is a consequence of the above fact $(\det f_*(1_X)) = [\mathrm{ord}_R \text{ coker } f^*]^2$ applied to the inclusion $f: Y^{S^1} \to Y$ together with a suitable analog of (vi).

Recent Developments

It has taken two years to arrange these notes in their present form. In the meantime they have motivated further results which justify the development of what is here and make the speculations both hidden and explicit (in section 9) in the notes more secure and worthwhile. Let us review the developments.

The fundamental problem which encompasses everything is: _Given is a G manifold Y. How can we construct all G manifolds X together with G maps $f:X \to Y$ such that $|f| : |X| \to |Y|$ is a homotopy equivalence; moreover, given such a map f what restrictions are imposed on the local G invariant of X and Y?_

The special case $G = 1$ has seen a vigorous and fruitful history for which there is a complete solution. There are three fundamental concepts involved: fiber homotopy equivalence of vector bundles, transversality and surgery (cobordism theory). Briefly one starts with a fiber homotopy equivalence $\omega : \xi \to \eta$ of vector bundles over Y and via a proper homotopy converts ω to a proper map $\theta : \xi \to \eta$ which is transverse to the zero section $Y \subset \eta$ (written $\theta \pitchfork Y$). Then $X = \theta^{-1}(Y)$ is a smooth manifold and $\theta | X = f: X \to Y$ is a degree one map (with some additional structure). The technique for converting f to a homotopy equivalence is surgery (normal cobordism).

The three concepts mentioned above have important generalizations for G arbitrary. They are quasi-equivalence of G vector bundles, G transversality and G normal cobordism.

1. <u>Quasi-Equivalence</u>: Let ξ and η be two G vector bundles (complex) over Y of the same dimension. A G map $\omega:\xi \to \eta$ which is proper, fiber preserving and degree one on fibers is called a quasi-equivalence. Define $\xi \leq \eta$ to mean there exists a G bundle θ and a quasi-equivalence $\omega:\xi \oplus \theta \to \eta \oplus \theta$.

When Y is a point Meyerhoff—Petrie in <u>Quasi-Equivalence of G modules</u> give necessary and sufficient conditions for $N \leq M$ when N and M are complex G modules (viewed as G bundles over a point). This has applications to general Y since $N \leq M$ implies $Y \times N \leq Y \times M$ as G vector bundles over Y. This remark and the results of the above paper give a vast supply of quasi-equivalent vector bundles over the $Y = P(\Omega)$ of Part II, §8 when we restricted attention to $N = t^p \oplus t^q$ and $M = t^1 \oplus {}^{pq}$.

2. G transversality. Petrie in G <u>transversality</u> has given an obstruction theory for the problem of constructing a proper G homotopy between $f:N \to M$ and a map θ transverse to an invariant submanifold Y of M which in particular applies to the case f is a quasi-equivalence of G vector bundles over Y. Roughly one can say that each isotropy group K of the action on N gives a sequence of obstructions

3 $$O_*(f,K) \in H^*(X^K/N(K), X_K/N(K), \pi_{*-1}(V(K)))$$

where f is properly G homotopic to f'

$$f'^K \pitchfork Y^K,\ X^K = (f'^K)^{-1}(Y^K),$$

$X_K = \bigcup_{H>K} X^H$, $N(K)$ is the normalizer of K in G and $V(K)$ is a manifold which depends on the representations of K on the normal fiber to Y in M at $y \in Y^K$ and the normal fiber to N^K in N at $x \in X^K$. If all obstructions vanish f is properly G homotopic to $\theta \pitchfork Y$ and we obtain a G manifold $X = \theta^{-1}(Y)$.

The obstruction theory is especially suitable for treating the case where f is a quasi-equivalence $f: Y \times N \to Y \times M$ induced from a quasi-equivalence of N to M. Taking $G = S^1$ $Y = P(\Omega)$, the above N and M, gives the manifolds $X(\omega)$ and map $\theta = \bar{F}$ of Part II, Proposition 8.2. Dejter, in his thesis, gives necessary and sufficient conditions on N, M and Ω so that a quasi-equivalence $f: P(\Omega) \times N \to P(\Omega) \times M$ be properly G homotopic a map transverse to $P(\Omega)$.

4. G Normal Cobordism. This concept is precisely defined in Part II, 9. It sufficies here to say that it is a technique for converting a G map $f: X \to Y$ to a G map $f': X' \to Y$ with $|f'|$ a homotopy equivalence. Petrie in Equivariant Quasi-Equivalence, Transversality and Normal Cobordism, Proceedings

T.C.M., Vancouver (1974) gives some algebraic obstructions to producing such an f' and discusses situations where the obstructions are adequate to solve the problem.

In general much remains to be done. There is an interesting setting where the present technology of G normal cobordism is probably sufficient to solve the problem of producing f'. This is the case of elementary isotropy on Y i.e. for each isotropy group H of the action on Y, $N(H)/H$ acts freely on Y^H.

These three concepts provide the constructive tools for treating the fundamental problem mentioned above. In particular they have been applied in Part II §8 and in the literature cited above to treat the study of the pseudo-free S^1 manifolds of Montgomery-Yang.

These ideas were originally developed to deal with the

Conjecture: (Petrie, Bull. A.M.S. Vol. 78, No. 2 (1972)): Let X be a closed S^1 manifold with $X^{S^1} \neq X$ which admits a homotopy equivalence $h : |X| \longrightarrow CP^n$. Then h preserves Pontrjagin classes.

Using I Theorem 6.19 and II §10 among other tools, Dejter has proved the conjecture for $n = 3$.

These applications together with the satisfactory status of quasi-equivalence and G transversality enhance the interest in II §9 towards developing invariant for the theory of G normal cobordism.

Much remains to be done on all these ideas as well as their

P A R T I

THE ALGEBRAIC TOOLS

1. Preliminary remarks on G spaces

Let $C(G)$ denote the category of compact G spaces. An object X of $C(G)$ can be viewed as a representation of the compact Lie Group G in the group $H(|X|)$ of homeomorphisms of the underlying compact topological space $|X|$ of X. We require that the map from $G \times X \longrightarrow X$ defined by $(g,x) \longrightarrow X(g)x$ be continuous in the product topology on G and $|X|$. A map $f : X \longrightarrow Y$ is a continuous map $|f| : |X| \longrightarrow |Y|$ which commutes with the action of G on $|X|$ and $|Y|$ defined by the representations X and Y. In greater detail, if $g \in G$, $X(g)$, and $Y(g)$ are homeomorphisms of $|X|$ and $|Y|$, and we require

$$|f| \circ X(g) = Y(g) \circ |f| .$$

In order to simplify notation, we use the same symbol X for the G space and its underlying topological space $|X|$. Also, we make the abreviations

$$gx \quad \text{or} \quad g \circ x \quad \text{for} \quad X(g)(x)$$

where $g \in G$, and $x \in X$.

The category $C(G)$ is a subcategory of the category $LC(G)$ of locally compact G spaces. Here are some important examples of objects in these categories:

E.1 <u>The point p</u>. This is the G space consisting of a single point on which G acts trivially, i.e., $gp = p$ for all $g \in G$.

E.2 The product $X \times Y$ $X, Y \in LC(G)$. Here $g \in G$ acts on $(x,y) \in X \times Y$ by $g(x,y) = (gx, gy)$.

E.3 Real (complex) G modules. Denote the real numbers by $\mathbb{R}$ and the complex numbers by $\mathbb{C}$. A real (complex) G module M consists of a real (complex) vector space $|M| = \mathbb{R}^n$ $(\mathbb{C}^n)$ for some n together with G acting through an orthogonal (unitary) representation of G, i.e., there is a homomorphism $M : G \longrightarrow O(n)$ $(U(n))$ such that for $m \in M$

$$gm = M(g)m$$

is the result of applying the orthogonal (unitary) matrix $M(g)$ to $m \in M$.

E.4 The unit disk $D(M)$ and unit sphere $S(M)$ are G subspaces of M in an obvious way.

E.5 Real (complex) G vector bundles over a G space. Let $X \in LC(G)$. Then $V \in LC(G)$ is a G vector bundle over X if there is a map $\pi_V : V \longrightarrow X$ in $LC(G)$ such that

(i) $|\pi_V| : |V| \longrightarrow |X|$ is the projection of a real (complex) vector bundle over $|X|$ [1].

(ii) For each $x \in X$, the map

$$g : |\pi_V|^{-1}(x) \longrightarrow |\pi_V|^{-1}(gx)$$

is real (complex) linear. (In fact we want this to preserve a Riemanian metric on the bundle $|V|$.)

Associated to a G vector bundle V, we have the unit disk bundle $D(V)$ and the unit sphere bundle which are also G spaces. Let

1.6 $$i_{S(V)} : S(V) \longrightarrow D(V)$$

denote the inclusion and

1.7 $$s_V : X \longrightarrow V$$

the zero section which maps each $x \in X$ to the zero of the vector space $\pi_V^{-1}(x)$

1.8 $$s_{D(V)} : X \longrightarrow D(V)$$

the zero section of the disk bundle.

Here are some useful examples of G vector bundles:

E.9 $X \times M = \hat{M}$. If X is any G space and M is a real (complex) G module, then $X \times M$ is a G vector bundle over X which we abbreviate by $\hat{M}$. The projection $\pi_{\hat{M}} : \hat{M} \longrightarrow X$ is defined by

$$\pi_{\hat{M}}(x,m) = x.$$

E.10 TX. If X is a smooth manifold with G acting smoothly, then its tangent bundle TX is a real G vector bundle over X. The action of G on TX is defined by

$$g \circ v = dg(v) \qquad v \in TX$$

and $dg : TX \longrightarrow TX$ is the differential of $g : X \longrightarrow X$ when $g \in G$.

We suppose given a cohomology theory h^* on $C(G)$ with values in the category $A(R)$ of Z_2 graded algebras over the ring

1.11 $$h^0(p) = R.$$

This means to each pair $(X,Y) \in C(G)$ with $Y \subset X$, we attach a Z_2 graded R algebra $h^*(X,Y)$ which is denoted by $h^*(X)$ when Y is empty and in that case has a unit $1 \in h^0(X)$. Specifically, $h^*(X,Y) = h^0(X,Y) \oplus h^1(X,Y)$. To each map $f : (X,Y) \longrightarrow (X',Y')$ we attach an R homomorphism

$$f^* : h^*(X',Y') \longrightarrow h^*(X,Y)$$

and for each inclusion $i : Y \longrightarrow X$ we attach an exact sequence

1.12 $$\cdots \longrightarrow h^k(X,Y) \xrightarrow{j^k} h^k(X) \xrightarrow{i^k} h^k(Y) \xrightarrow{\delta} h^{k+1}(X,Y)$$
$$\xrightarrow{j^{k+1}} h^{k+1}(X) \xrightarrow{i^{k+1}} h^{k+1}(Y) \xrightarrow{\delta} h^{k+2}(X,Y) = h^k(X,Y) \longrightarrow$$

indexed on the integers mod 2. We abbreviate this to an exact triangle

1.13 $$\begin{array}{ccc} h^*(X,Y) & \xrightarrow{j^*} & h^*(X) \\ {\scriptstyle\delta}\nwarrow & & \swarrow{\scriptstyle i^*} \\ & h^*(Y) & \end{array}$$

In the applications the functor h^* will be $K_G^*(\ )_S$, equivariant complex K theory localized at some set S of prime ideals of $R(G)$, or some close variant of $K_G^*(\ )$. Because of this we feel free to use most of the properties of this functor K_G^* as properties of h^*.

1.14 Note: h^* is a <u>contravariant</u> functor on the category $C(G)$ with values in $A(R)$.

Following [3], we extend the cohomology theory h^* from $C(G)$ to $LC(G)$ as follows: If $X \in LC(G)$, let X^+ denote its one point compactification with $+$ denoting the point at infinity. The action of G on X^+ is defined by

(i) $gx = X^+(g)(x) = X(g)(x)$ if $x \in X \subset X^+$

(ii) $g+ = +$.

Note that $+ \subset X^+$ is a G invariant subspace, so we define

1.15 $$h^*(X,Y) = h^*(X^+,Y^+)$$

when $Y \subset X$ is a closed subspace. If Y is the empty set ϕ then $\phi^+ = +$. With this definition

$$h^*(X) = h^*(X^+,+)$$

is an R algebra with unit iff X is compact.

Now suppose that $f : X \longrightarrow Y$ is a map of $LC(G)$ and that f is proper. This means that if $K \subset Y$ is compact, $f^{-1}(K)$ is a compact subset of X. In this case there is a map

$$f^+ : X^+ \longrightarrow Y^+ \quad \text{in} \quad C(G)$$

defined by

$$f^+(x) = f(x) \qquad x \in X$$

$$f^+(+) = +$$

and by definition $f^* : h^*(Y) \longrightarrow h^*(X)$ is the map $(f^+)^* : h^*(Y^+,+) \longrightarrow h^*(X^+,+)$. Hence on the category of locally compact G spaces and proper maps h^* is a cohomology theory, in particular a <u>contravariant</u> functor for <u>proper</u> maps.

Let $Y \subset X$ be an <u>open</u> subset. Let $i : Y \longrightarrow X$ denote the inclusion. Since Y is open in X, there is a map

$$i_+ : X^+ \longrightarrow Y^+$$

defined by

$$i_+(y) = y \qquad y \in Y \subset X^+$$

$$i_+(y) = + \qquad y \in X^+ - Y$$

Then i_+ is a map in $C(G)$ and we can define

1.16 $$i_* : h^*(Y) \longrightarrow h^*(X)$$

to be $$(i_+)^* : h^*(Y^+,+) \longrightarrow h^*(X^+,+)$$

1.17 Note: h^* is a <u>covariant</u> functor for <u>open</u> maps.

Suppose that $Y \subset X$ is a closed G subspace of X. Let I_0 denote the half open interval $[0,1)$ with trivial action of G and let

1.18 $$X/_Y = X \cup Y \times I_0$$

denote the G space obtained from the disjoint union of X and $Y \times I_0$ by identifying $y \in Y \subset X$ with $(y,0) \in Y \times I_0$. Note that the inclusion $X - Y \subset X/_Y$ is an open imbedding denoted by

1.19 $$\lambda : X - Y \longrightarrow X/_Y$$

Observe that the definitions, excision, and 1.15 imply that

1.20 $$h^*(X-Y) = h^*(X^+, Y^+) = h^*(X,Y)$$

when Y is a closed subspace of X and

1.21 $$h^*(X,Y) = h^*(X/_Y) .$$

From 1.20 it follows that when Y is a closed subset of the locally compact G space X, we have an exact triangle of the form:

1.22 $$\begin{array}{ccc} h^*(X-Y) & \xrightarrow{i_*} & h^*(X) \\ & {}_{\delta}\nwarrow \quad \swarrow_{j^*} & \\ & h^*(Y) & \end{array}$$

where $i : X\text{-}Y \longrightarrow X$ is the <u>open</u> inclusion and $j : Y \longrightarrow X$ is the inclusion of the <u>closed</u> subspace Y.

From 1.20 and 1.21, we see that $h^*({}^X/_Y)$ and $h^*(X\text{-}Y)$ are isomorphic. In fact

<u>Proposition 1.23</u> $\lambda_* : h^*(X\text{-}Y) \longrightarrow h^*({}^X/_Y)$ <u>is an isomorphism</u>.

Proof: Observe that $X\text{-}Y = {}^X/_Y - Y\times I_0$ and $Y\times I_0$ is a closed G subspace of ${}^X/_Y$. Applying 1.22 with appropriate substitutions gives the exact triangle

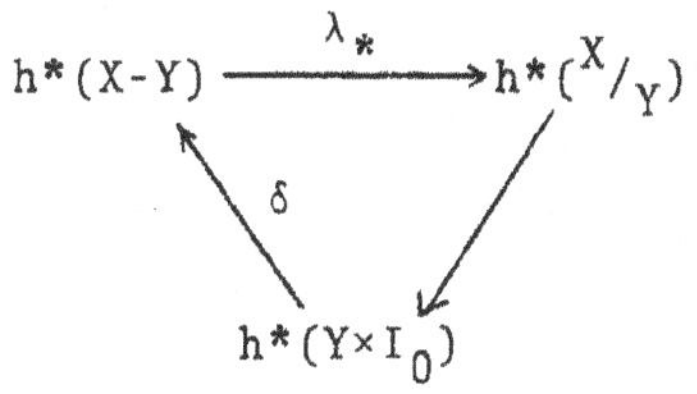

But $h^*(Y\times I_0) = h^*((Y\times I_0)^+, +) = h^*(C_r(Y^+),+) = 0$. Here $C_r(Y^+)$ is the space obtained from $Y^+ \times [0,1]$ by identifying every point of $Y^+ \times 1 \cup + \times [0,1]$ with $+ \times 1 = +$. Now $C_r(Y^+)$ and $+$ are compact G spaces and the inclusion of $+$ in $C_r(Y^+)$ is a G homotopy equivalence. This shows that $h^*(C_r(Y^+),+) = 0$.

Using this proposition we can give a description of the coboundary homomorphism δ.

Let $\dot{I}$ denote the interior of the unit interval, i.e., $\dot{I} = (0,1)$ with trivial action of G and let

1.24 $$\hat{\delta} : Y\times\dot{I} \longrightarrow {}^{X}/_{Y}$$

denote the open inclusion.

Definition 1.25. The coboundary homomorphism $\delta : h^*(Y) \longrightarrow h^{*+1}(X-Y)$ is the composition:

$$h^*(Y) = h^{*+1}(Y\times\dot{I}) \xrightarrow{\hat{\delta}_*} h^{*+1}({}^{X}/_{Y}) \xrightarrow{(\lambda_*)^{-1}} h^{*+1}(X-Y).$$

Remark: The equality $h^{*+1}(Y\times\dot{I}) = h^*(Y)$ is the suspension isomorphism for locally compact spaces. In fact, for the theory K_G^*, one <u>defines</u>

$$K_G^1(X) = K_G^0(X\times\dot{I}). \quad \text{See [3].}$$

There is an important instance when we can give a description of the homomorphism $(\lambda_*)^{-1}$ used to define the coboundry homomorphism. This is the case when Y is the boundry of the smooth G manifold X. In particular, X and Y are elements of the category of smooth G manifolds $D(G)$. This is the subcategory of $LC(G)$ defined by requiring an element $X \in D(G)$ to be a representation

1.24 $$X : G \longrightarrow \mathrm{Diff}\ (|X|)$$

of G into the group of diffeomorphisms of the smooth manifold $|X|$. Moreover, we require the map

$$\mu : G\times|X| \longrightarrow |X|$$

defined by $\mu(g,x) = X(g)(x)$ to be smooth.

Suppose that $X \in D(G)$ is a smooth G manifold with boundry ∂X. Then ∂X is a smooth G submanifold of X.

Proposition 1.25 <u>There is a G homeomorphism</u>

$$\omega : X/_{\partial X} \longrightarrow X - \partial X = \dot{X}.$$

Proof: This is an easy consequence of the smooth G collaring theorem.

Collaring Theorem 1.26 [6] <u>Suppose that G is a compact Lie Group acting smoothly on the locally compact smooth manifold X with boundry ∂X. Then there exists a G homeomorphism c of $\partial X \times I$ (G acts trivially on I) onto a neighborhood of ∂X in X with</u>

$$c(x,0) = x \qquad x \in \partial X.$$

If X and $Y = \partial X$ are as in the Collaring Theorem then the coboundry homomorphism

$$\delta : h^*(\partial X) \longrightarrow h^*(X,\partial X) = h^*(\dot{X})$$

is defined by

$$1.26 \qquad h^*(\partial X) = h^{*+1}(\partial X \times \dot{I}) \xrightarrow{\hat{\delta}} h^{*+1}(X/_{\partial X}) \xrightarrow{\omega^*} h^{*+1}(\dot{X}).$$

That is $\omega^* = (\lambda_*)^{-1}$.

For example, if V is a real G module and $X = D(V)$ then $X/_{\partial X} = D(V)/S(V)$ can be identified with

$$D_2(V) = \{v \in V \mid \|v\| < 2\}$$

and it is easy to write down a G homeomorphism of $D_2(V)$ with $D(\dot{V}) = \{v \in V \mid \|v\| < 1\}$.

Actually it is the category $D(G)$ and $D_C(G)$ the subcategory of compact smooth G manifolds that we want to study. On the category $D_C(G)$ we can define a homology theory h_* dual to the cohomology theory h^*. In order to do this we need a method of associating in a canonical manner to each compact smooth G submanifold Y of X a smooth compact G submanifold N containing Y as a G deformation retract and having the same dimension as X. Such a manifold N is called a closed G tubular neighborhood of Y in X. When Y is a compact manifold without boundry in the interior of X, the existence and uniqueness of G tubular neighborhood is discussed in [6]. We can even drop the assumption that Y has no boundry although this is not discussed in [6].

<u>We will only be concerned with submanifolds</u> $Y \subset X$ <u>whose interior</u> $\dot{Y}$ <u>does not intersect the boundary of</u> X. If $Y \subset \partial X \subset X$, then $Y \subset X/\partial X$ which is homeomorphic to $\dot{X}$. Thus Y has a G tubular neighborhood in $\dot{X}$. We call this a G tubular neighborhood of Y in X. In particular, this devise gives a G tubular neighborhood of ∂X in X.

The homology functor on the category $D_C(G)$ is roughly the composition of cohomology with the tangent space functor, i.e., $h_* = h^* \circ T$. More precisely let $Y \subset X$, $Y, X \in D_C(G)$. First assume that dimension Y = dimension X and let X denote the interior of X.

Definition 1.27

$$h_i(X,Y) = h^i(T(\dot{X}-\dot{Y}))$$

where $T(\dot{X}-\dot{Y})$ is the tangent space of $\dot{X}-\dot{Y}$.

If $Y \subset X$ is a G submanifold of dimension less than X, let $N \subset X$ denote a closed tubular neighborhood of Y in X. See [6], pp. 303-312.

Definition 1.27′

$$h_i(X,Y) = h^i(T(\dot{X}-\dot{N})).$$

Note that if Y is empty and X is closed (i.e., X is compact and $\partial X = \phi$)

1.28 $$h_i(X) = h^i(TX) \text{ ; more generally}$$

$$h_i(X,\partial X) = h^i(TX).$$

In order to make h_* a covariant functor for maps in the category $D_C(G)$ we need a Thom isomorphism in cohomology for complex G vector bundles. We take this opportunity to digress on the relationship between G vector bundles and cohomology.

When $h^* = K_G^*$ and V is a <u>complex</u> G vector bundle over $X \in C(G)$, there is an element $\lambda_V \in K_G^0(V)$ constructed from the exterior algebra of V. This class generates $K_G^*(V)$ as a free $K_G^*(X)$ module. Since our theory h^* is either K_G^* or a simple variant, there is an element in $h^0(V)$ with similar properties, in fact we postulate the existence of

1.29 $$\lambda_V \in h^0(V)$$

such that the Thom homomorphism

1.30 $$\psi^V : h^*(X) \longrightarrow h^*(V)$$

defined by $\psi^V(x) = \lambda_V \cdot x$ is an isomorphism. Here dot denotes multiplication of $\lambda_V \in h^0(V)$ by $x \in h^*(X)$. This of course uses the fact that $h^*(V)$ is an $h^*(X)$ module. It is not difficult to see that $h^*(V)$ is an $h^*(X)$ module once one observes that $h^*(V)$ is $h^*(D(V),S(V))$ which is a module over $h^*(D(V))$ which in turn is a module over $h^*(X)$ via the ring homomorphism $\pi^*_{D(V)}; \pi_{D(V)}: D(V) \longrightarrow X$.

Before discussing the functorial properties of λ_V and ψ^V we must mention two important methods of constructing new G vector bundles from existing G vector bundles.

1.31 The induced bundle. Let $f : X \longrightarrow Y$ be a map in $LC(G)$ and let V be a G vector bundle, real or complex, over Y. Let $f^*V \subset X \times V$ denote the set of points (x,v) such that $f(x) = \pi_V(v)$. Then f^*V is a G invariant subspace of $X \times V$. Define $\pi_{f^*(V)} : f^*V \longrightarrow X$ by $\pi_{f^*(V)}(x,v) = x$. Then f^*V with projection $\pi_{f^*(V)}$ is a G vector bundle over X called the bundle induced by f or the induced bundle.

As an application of 1.31, we mention the Whitney sum. Suppose V and W are G vector bundles over X.

Then $V\times W$ is in an obvious way a G vector bundle over $X\times X$. Let $\Delta : X \longrightarrow X\times X$ be the map defined by $\Delta(x) = (x,x)$ then

1.32 Define $V\oplus W = \Delta^*(V\times W)$. This is the Whitney sum of V and W.

When $i : Y \longrightarrow X$ is an inclusion and V is a G vector bundle over X we set $i^*V = V|Y$.

As for the functorial property of λ_V and ψ^V (in the noncompact case), we suppose λ_V and ψ^V are natural with respect to maps. This means that if $f : X \longrightarrow Y$ is a map in $LC(G)$ and V is a G vector bundle over Y,

1.33 $$f^*\lambda_V = \lambda_{f^*(V)}$$

and

1.34 $$f^*\psi^V(y) = \psi^{f^*(V)}(f^*(y)), \qquad y \in h^*(Y)$$

When $X \in C(G)$ and V is a <u>complex</u> G vector bundle over X, we define a class $\lambda_{-1}(V) \in h^0(X)$ by

1.35 $$\lambda_{-1}(V) = s_V^*\lambda_V$$

and note

1.36 $$s_V^*\psi^V(x) = s_V^*(\lambda_V\cdot x) = \lambda_{-1}(V)\cdot x$$

when $x \in h^*(X)$. This is a consequence of the naturality of cross products discussed in §3. See also [5].

We suppose that when V and W are two complex G vector bundles over $X \in C(G)$ then

1.37 $$\lambda_{-1}(V \oplus W) = \lambda_{-1}(V) \cdot \lambda_{-1}(W).$$

We are now in position to describe the way in which h_* is a covariant functor on $D_C(G)$. Suppose first that X and Y are in $D_C(G)$ with $X \subset Y$. Let i denote the inclusion of X in Y. We want to define an R homomorphism

$$i_* : h_*(X) \longrightarrow h_*(Y).$$

Let N denote a G tubular neighborhood of X in Y. Then $\dot{N}$, the interior of N, may be identified with the G normal bundle of $\dot{X}$ in $\dot{Y}$. (This is a real G vector bundle over $\dot{X}$.) The tangent bundle of X, TX, is a G submanifold of TY and the tubular neighborhood $T\dot{N}$ of $T\dot{X}$ in $T\dot{Y}$ can be identified with the vector bundle

$$\pi^*_{T\dot{X}}(\dot{N} \oplus i\dot{N}) = \pi^*_{T\dot{X}}(\dot{N} \underset{R}{\otimes} C)$$

over $T\dot{X}$. Here $\pi_{T\dot{X}} : T\dot{X} \longrightarrow \dot{X}$ is the projection of the real G vector bundle $T\dot{X}$ on $\dot{X}$.

Since $\pi^*_{T\dot{X}}(\dot{N} \otimes C) = \dot{N}_c$ is a complex G vector bundle over $T\dot{X}$, we have the Thom isomorphism

$$\psi^{\dot{N}_c} : h^*(T\dot{X}) \longrightarrow h^*(T\dot{N}).$$

We shall agree that $\pi^*_{T\dot{X}}(\dot{N})$ corresponds to points of Y and $\pi^*_{T\dot{X}}(i\dot{N})$ to tangent vectors of Y.

Since $T\dot{N}$ is open in $T\dot{Y}$ we have the homomorphism $k_* : h^*(T\dot{N}) \longrightarrow h^*(T\dot{Y})$ induced by the open inclusion $k : T\dot{N} \longrightarrow T\dot{Y}$.

Define $i_* : h_*(X) \longrightarrow h_*(Y)$ as the composition

$$h_*(X) = h^*(T\dot{X}) \xrightarrow{\psi^{\dot{N}_C}} h^*(T\dot{N}) \xrightarrow{k_*} h^*(T\dot{Y}) = h_*(Y).$$

In particular, if we denote by $Ti : T\dot{X} \longrightarrow T\dot{Y}$ the inclusion defined by $i : X \longrightarrow Y$ and set $(Ti)_* = k_*\psi^{\dot{N}_C}$, then i_* is defined by $(Ti)_*$ and if $\partial X = \phi$,

1.38 $$(Ti)^*(Ti)_*(x) = \pi^*_{TX}(\lambda_{-1}(N\otimes C))\cdot x$$

$x \in h^*(TX)$ where $\pi_{TX} : TX \longrightarrow X$ is the projection. See [5], p. 497 (3.1).

In a similar manner we can define $i_* : h_*(X,A) \longrightarrow h_*(Y,B)$ when $i : (X,A) \longrightarrow (Y,B)$ is an inclusion of compact smooth G manifolds.

In the above manner, we have defined i_* when $i : X \longrightarrow Y$ is an inclusion. It remains to define i_* for arbitrary maps in $D_C(G)$.

A Theorem of Mostow [5], p. 111 asserts that if $X \in D_C(G)$, there is a real G module M and a smooth G imbedding

$$\alpha : X \longrightarrow M.$$

Let $f : X \longrightarrow Y$ be a map in $D_C(G)$.

Define

$$f' : X \longrightarrow Y\times M$$

by

$$f'(x) = (f(x), \alpha(x)).$$

Then f' is an inclusion, so

$$(Tf')_* : h^*(T\dot{X}) \longrightarrow h^*(T(\dot{Y}\times M))$$

is defined. But $T(\dot{Y}\times M) = T\dot{Y} \times TM$ and $TM = M\otimes_R C$ is a complex G module, so $T\dot{Y} \times TM$ is a complex G vector bundle over $T\dot{Y}$ namely, $M\hat{\otimes}_R C$. See E.9. We define $f_* : h_*(X) \longrightarrow h_*(Y)$ by the composition

1.39 $$h_*(X) = h^*(T\dot{X}) \xrightarrow{(Tf')_*} h^*(M\hat{\otimes}_R C) \xrightarrow{\left(\psi^{M\hat{\otimes}_R C}\right)^{-1}} h^*(T\dot{Y}) = h_*(Y).$$

In particular, for a pair (X,A) of elements of $D_C(G)$ with $A\subset X$ and $\dim A = \dim X$ we have an exact triangle

1.40 $$\begin{array}{ccc} h_*(A) & \xrightarrow{i_*} & h_*(X) \\ & {\scriptstyle\partial}\nwarrow \quad \swarrow{\scriptstyle j_*} & \\ & h_*(X,A) & \end{array}$$

induced by this exact sequence in cohomology

1.41
$$\begin{array}{ccc} h^*(T\dot{A}) & \xrightarrow{(Ti)_*} & h^*(T\dot{X}) \\ & \nwarrow \delta \qquad \alpha^* \swarrow & \\ & h^*(T(\dot{X}-\dot{A})) & \end{array}$$

Here $T(\dot{X}-\dot{A}) \xrightarrow{\alpha} T\dot{X}$ is the inclusion of the <u>closed</u> G subspace $T(\dot{X}-\dot{A})$. The homomorphism δ is defined by observing that

1.42 $$h^*(T\dot{A}) = h^*(T\dot{X}-T(\dot{X}-\dot{A})) = h^*(T\dot{X},T(\dot{X}-\dot{A}))$$

In particular, we call attention to the fact that the homomorphism ∂ in homology is defined via the homomorphism δ in cohomology.

Remark 1.43 The exact sequence 1.40 exists without the assumption $\dim A = \dim X$. This assumption is removed by replacing A by a closed tubular neighborhood of A in X.

Thus to each $f : X \longrightarrow Y$ in $D_C(G)$, we have defined $f_* : h_*(X) \longrightarrow h_*(Y)$. In particular, for any $X \in D_C(G)$ there is a unique map

1.44 $$\rho^X : X \longrightarrow P$$

The R homomorphism

1.45 $$\rho_0{}^X : h_0(X) \longrightarrow h_0(p) = R$$

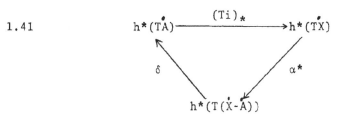

is of particular importance. We call this the index homomorphism and distinguish it by setting

$$\text{1.46} \qquad \mathrm{Id}_R^X = \rho_0^X .$$

The index homomorphism satisfies the following two important properties:

Proposition 1.46 *If* $f : X \longrightarrow Y$ *is a map in* $D_C(G)$, *the following diagram is commutative*

$$\begin{array}{ccc} h_0(X) & \xrightarrow{\quad f_0 \quad} & h_0(Y) \\ & \mathrm{Id}_R^X \searrow \quad \swarrow \mathrm{Id}_R^Y & \\ & R & \end{array}$$

Proposition 1.47 $\mathrm{Id}_R^p : R = h_0(p) \longrightarrow R$ *is the identity.*

Remarks 1.48 Let X and Y be in $LC(G)$ and $f : X \to Y$. Then $h^*(X)$ is an $h^*(Y)$ module. This is clear if X and Y are compact. In the general case note that the map $f : X \longrightarrow X \times Y$ defined by

$$f(x) = (x, f(x))$$

is proper. Then

$$h^*(X) \otimes h^*(Y) \xrightarrow{\ \times\ } h^*(X \times Y) \xrightarrow{\ f^*\ } h^*(X)$$

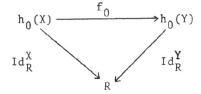

defines $h^*(X)$ as an $h^*(Y)$ module. Here $\times$ denotes the cross product discussed in Section 5.

Now assume that Y is compact and that V is a G vector bundle over X which is induced from a G vector bundle V' over Y. If X is not compact, then $\lambda_{-1}(V)$ does not exist as an element of $h^*(X)$; none-the-less, we can define $\lambda_{-1}(V)\cdot\mu \in h^*(X)$ for any $\mu \in h^*(X)$, i.e., we define multiplication by $\lambda_{-1}(V)$ in $h^*(X)$. By definition

$$\lambda_{-1}(V)\cdot\mu = \lambda_{-1}(V')\cdot\mu.$$

The right hand side of this expression is defined because $\lambda_{-1}(V') \in h^*(Y)$ and $h^*(X)$ is an $h^*(Y)$ module.

In like manner the class λ_V is not defined as an element of $h^*(V)$. We can still define $\mu \longrightarrow \lambda_V u \in h^*(V)$ for $u \in h^*(X)$ by $\lambda_V\cdot\mu = \lambda_{V'}\cdot\mu$ where $\lambda_{V'}\cdot\mu$ denotes the image of $\lambda_{V'} \otimes \mu \in h^*(V') \otimes h^*(X)$ under the homomorphism

$$h^*(V') \otimes h^*(X) \xrightarrow{\times} h^*(V'\times X) \xrightarrow{g^*} h(V)$$

where $g : V \longrightarrow X\times V'$ is the proper map defined by

$$g(v) = (\pi_V(v),\ f^*(v))$$

and $f^* : V \longrightarrow V'$ is the G vector bundle map provided by $V = f^*(V')$.

We require that the homomorphism $\psi^V : h^*(X) \longrightarrow h^*(V)$ defined by $\psi^V(u) = \lambda_V\cdot\mu$ be an isomorphism and

$$(s_V)^*\psi^V(\mu) = \lambda_{-1}(V)\cdot\mu.$$

2 Structure of smooth G manifolds in terms of handles

If Γ is a subgroup ot G and $X \in LC(\Gamma)$, we can form a space

$$G\times_{\Gamma}X \in LC(G)$$

in the following manner: Let Γ act on $G\times X$ by the rule

$$\gamma(g,x) = (g\gamma^{-1},\gamma x) \qquad \gamma \in \Gamma, \quad g \in G, \quad x \in X.$$

Then $G\times_{\Gamma}X$ is defined to be the orbit space of $G\times X$ by this action of Γ. The point of $G\times_{\Gamma}X$ determined by $(g,x) \in G\times X$ is denoted by $[g,x]$. The action of G on $G\times_{\Gamma}X$ is defined by

$$g[g',x] = [gg',x] \qquad g,g' \in G, \quad x \in X$$

There are two important special cases of this construction. If $X=p$, then

$$G\times_{\Gamma}p = {}^{G}/_{\Gamma}$$

is the space of left cosets of G by Γ with G acting by left translation. If $X=V$ is a real (complex) Γ module then the G space $G\times_{\Gamma}V$ is a real (complex) G vector bundle over $G\times_{\Gamma}P = {}^{G}/_{\Gamma}$. We abreviate this real (complex) G vector bundle by V. The projection π_V of V on $G\times_{\Gamma}P = {}^{G}/_{\Gamma}$ is defined by

$$\pi_V[g,v] = [g] \in {}^{G}/_{\Gamma}$$

where $[g]$ denotes the left coset defined by $g \in G$.

If V and W are two real (complex) Γ modules define

$$D'(V \oplus W) = \{(v,w) \in V \oplus W \mid \|v\| \le 1, \ \|w\| \le 1\}$$

$$S_V(V \oplus W) = \{(v,w) \in V \oplus W \mid \|v\| = 1, \ \|w\| \le 1\}.$$

Definition 2.1: A G handle H with isotropy group Γ is an $H \in D_C(G)$ which is G homeomorphic to $G \times_\Gamma D'(V)$ for some real Γ module.

Definition 2.2 Let Z and X be elements of $D_C(G)$ of the same dimension. We say that Z is obtained from X by attaching a handle H with isotropy group Γ and index V if:

(i) V is a real Γ module and there is a real Γ module W such that $H = G \times_\Gamma D'(V \oplus W)$.

(ii) $Z = X \cup H$.

(iii) $H \cap X = H \cap \partial X = G \times_\Gamma S_V(V \oplus W)$.

The importance of G handles is that they are the building blocks for all compact G manifolds. Specifically, we have the following structure theorem:

Theorem 2.3 [14] Let X be a closed G manifold (compact without boundry) in $D_C(G)$. Let $Y \subset X$ be a closed invariant submanifold with G tubular neighborhood N (the unit disk bundle of the tubular neighborhood N of Y in X is $D(N$ Then there is a decreasing filtration $X = X_0 \supset X_1 \cdots \supset X_n = D(N)$ such that each X_i is obtained from X_{i+1} by attaching a G handle H_i with isotropy group Γ_i and Γ_i is the isotropy group of a point of $X-D(N)$.

Remark 2.4 The isotropy group of a point $x \in X$ is the subgroup of G defined as $\{g \in G \mid gx = x\}$.

Remark 2.5 Suppose that $G = S^1$ and $Y = X^{S^1}$. Then the isotropy subgroups Γ_i occuring in the handle decomposition of X by adding G handles to $D(N)$ are proper closed subgroups of S^1 so are finite cyclic.

We see from Theorem 2.3, that real Γ modules for subgroups $\Gamma \subset G$ play an important role in the structure of G manifolds. In particular, for the group $G = S^1$ we describe all real Γ when Γ is finite cyclic. This is very easy. Let $h \in \Gamma$ be a generator. Every real Γ module is a direct sum of real Γ modules of the following three types:

(1) R^1; where R^1 denotes the one dimensional R vector space on which Γ acts trivially.

(2) R_-; where R_- denotes the one dimensional R vector space on which h acts by multiplication by -1.

(3) C'_ℓ where C'_ℓ is the real two dimensional vector space R^2 and h acts via the matrix

$$\begin{pmatrix} \cos\theta & \sin\theta \\ -\sin\theta, & \cos\theta \end{pmatrix} \qquad \begin{aligned} \theta &= \frac{2\pi\ell}{\gamma} \\ \gamma &= \text{order of } \Gamma \end{aligned}$$

Clearly C'_ℓ is the underlying real Γ module defined by the complex Γ module C_ℓ where $|C_\ell| = C$ and h acts on $|C_\ell|$ by multiplication by $\exp \frac{2\pi i\ell}{\gamma}$. Also $R^1 \oplus R^1$ is the underlying real Γ module of the complex Γ module C where Γ acts trivially on C and $R_- \oplus R_-$ is the underlying real Γ module of the complex Γ module C_- on which Γ acts by multiplication by -1. This elementary discussion shows that

Remark 2.6 If V is any real Γ module where Γ is finite cyclic then $V \oplus V = V \times V$ is the underlying real Γ module of a complex Γ module $V^2 = V \otimes_{\mathbb{R}} \mathbb{C}$.

The complex 1 dimensional S^1 module M_n defined for each integer n by

$$|M_n| = C, \qquad t \circ z = t^n \cdot z \qquad t \in S^1, \quad z \in C$$

where dot denotes complex multiplication is especially important. Every complex S^1 module is a direct sum of the modules $\{M_n, n \in Z\}$. The element of the complex representation ring of the circle, $R(S^1) = Z[t,t^{-1}]$, represented by M_n is t^n.

Proposition 2.7 <u>Let $\Gamma \subset S^1$ denote the cyclic group of order γ. Then as S^1 manifolds</u>

$$S^1/\Gamma = S^1 \times_\Gamma P = S(M_\gamma).$$

Proof: Define S^1 maps ρ and ψ between S^1/Γ and $S(M_\gamma)$ by

$$\rho(t) = [t^{1/\gamma}] \in S^1/\Gamma \qquad \text{for } t \in S(M_\gamma)$$

$$\psi(t) = t^\gamma \in S(M_\gamma) \qquad \text{for } t \in S^1/\Gamma$$

Then $\psi\rho = 1_{S(M_\gamma)}$ and $\rho\psi = 1_{S^1/\Gamma}$ are the identity maps.

For future use we need the following basic facts:

(2.8) Let V be the complex S^1 module of dimension n defined by

$$V = t^{\lambda_1} + \cdots t^{\lambda_n} \in Z[t,t^{-1}] = R(S^1)$$

where the λ_i's are integers. Define $\lambda_{-1}(V) \in R(S^1)$ by $\lambda_{-1}(V) = \sum(-1)^i\lambda^i(V)$ where $\lambda^i(V)$ is the ith exterior power of V. This is the operation λ_{-1} on $R(S^1) = K^0_{S^1}(p)$ defined by 1.35. At any rate an explicit formula for $\lambda_{-1}(V)$ is

$$\lambda_{-1}(V) = \prod_{i=1}^{n}(1-t^{\lambda_i}) \in Z[t,t^{-1}]$$

(2.9) Let $\Gamma \subset S^1$ be the cyclic group of order γ. Then the complex representation ring of Γ is given by

$$R(\Gamma) = R(S^1)/(\lambda_{-1}(M_\gamma)) = Z[t]/(1-t^\gamma)$$

We also need to know the cohomology of the unit sphere $S(V)$ of the complex S^1 module V when $h^* = K^*_{S^1}(\ )\otimes Q$ (See §0). This is easy to determine from the exact sequence of the pair $(D(V),S(V))$ and the Thom isomorphism. See [1], Corollary 2.7b, page 104. Since the argument is so easy, we reproduce it.

Proposition 2.10 Let $V = t^{\lambda_1} + \cdots + t^{\lambda_u}$ $\lambda_i \neq 0$ be the indicated complex S^1 module. Then $\lambda_{-1}(V) = \prod_{i=1}^{n}(1-t^{\lambda_i})$ and

$$h^0(S(V)) = {}^{R}\Big/\Big(\prod_{i=1}^{n}(1-t^{\lambda_i})\Big)$$

$$h^1(S(V)) = 0$$

Proof: V is S^1 homeomorphic to $D(V) - S(V)$. Let j denote the resulting open embedding of V in $D(V)$. Then the exact sequence of the pair $(D(V), S(V))$ becomes

2.11
$$\begin{array}{ccc} h^*(V) & \xrightarrow{\;j_*\;} & h^*(D(V)) \\ {\scriptstyle \delta}\nwarrow & & \swarrow{\scriptstyle i^*_{S(V)}} \\ & h^*(S(V)) & \end{array}$$

Replace $h^*(V)$ by $h^*(X)$ using the Thom isomorphism ψ^V and replace $h^*(D(V))$ by $h^*(X)$ using the isomorphism $s^*_{D(V)}$. Then 2.11 becomes

2.12
$$\begin{array}{ccc} h^*(X) & \xrightarrow{\;\phi\;} & h^*(X) \\ {\scriptstyle \delta'}\nwarrow & & \swarrow{\scriptstyle i^*_{S(V)}\pi^*_{D(V)}} \\ & h^*(S(V)) & \end{array}$$

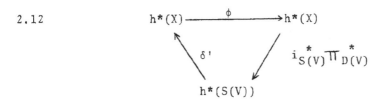

where $\phi = s^*_{D(V)} j_*$ is multiplication by $\lambda_{-1}(V)$. This uses the fact that $s^*_{D(V)} j_* = s^*_V$ and $s^*_V \psi^V(x) = \lambda_{-1}(V)\cdot x$ for $x \in h^*(X)$. Since $h^*(S(V)) = K^*_{S^1}(S(V))\otimes Q$ and $\lambda_{-1}(V) = \Pi(1-t^{\lambda_i})$ is nonzero in $R = Q[t,t^{-1}]$, the result follows from 2.12.

Remark 2.13 Let Z, X, V and W be as in Definition 2.2. Observe that the inclusion $(D(V),S(V)) \subset (D'(V\oplus W),S_V(V\oplus W))$ is a Γ homotopy equivalence of pairs and

$(S^1\times_\Gamma D'(V\oplus W),\ S^1\times_\Gamma S_V(V\oplus W)) = (H,H\cap X) \subset (Z,X)$ is an S^1 homotopy equivalence and

$$h^*(Z,X) = h^*(S^1\times_\Gamma D(V),S^1\times_\Gamma S(V)).$$

3. Multiplicative properties of $h^*(\ )$

We assume that our cohomology theory is equipped with products defined by a cross product homomorphism in $h^0(\ )$. This means the following: Let (X,A), (Y,B) be in LC(G). Define $(X,A) \times (Y,B) = (X\times Y,\ X\times B \cup A\times Y)$. Assume that

3.1 There is a cross product homomorphism

$$h^0(X,A) \underset{R}{\otimes} h^0(Y,B) \xrightarrow{\ \times\ } h^0((X,A) \times (Y,B))$$

which is natural with respect to maps.

3.2 There is a natural isomorphism

$$S : h^0\ (X,A) \xrightarrow{\ \cong\ } h^0(X,A)\times R^2)$$

compatible with cross products, i.e., we have a commutative diagram:

$$\begin{array}{ccc} h^0(X,A) \underset{R}{\otimes} h^0\ (Y,B) & \xrightarrow{\ \times\ } & h^0((X,A) \times (Y,B)) \\ \downarrow 1\otimes S & & \downarrow S \\ h^0(X,A) \underset{R}{\otimes} h^0((Y,B)\times R^2) & \xrightarrow{\ \times\ } & h^0((X,A) \times (Y,B)\times R^2) \end{array}$$

3.3 $h^i(X) = h^0(X\times R^i)$ and the cross product homomorphism in the cohomology theory h^* is defined by

$$\begin{array}{ccccc}
h^i(X,A) \underset{R}{\otimes} h^j(Y,B) & \xrightarrow{\ \times\ } & h^{i+j}((X,A) \times (Y,B)) & & \\
\| & & & \searrow{\scriptstyle \cong} & \\
h^0((X,A)\times R^i) \underset{R}{\otimes} h^0((Y,B)\times R^j) & \longrightarrow & h^0((X,A)\times R^i \times (Y,B)\times R^j) & \xrightarrow{\ T^*\ } & *
\end{array}$$

$* = h^0((X,A) \times (Y,B) \times R^{i+j})$. Here

$$T : (X,A) \times (Y,B) \times R^i \times R^j \longrightarrow (X,A) \times R^i \times (Y,B) \times R^j$$

is defined by $T(x,y,r_i,r_j) = (x,r_i,y,r_j)$ when $x \in X$, $y \in Y$, $r_i \in R^i$, $r_j \in R^j$.

3.4 Using cross product, we can define multiplication denoted

$$h^*(X,A) \underset{R}{\otimes} h^*(X,B) \longrightarrow h^*(X,A \cup B)$$

by $a \cdot b = \Delta^*(a \times b)$ where

$$\Delta : X \longrightarrow X\times X \quad \text{is} \quad \Delta(\mu) = (\mu,\mu) \in X\times X.$$

3.5 With this definition of cross product and multiplication, the two are related by

(*) $\quad a \times b = \pi_X^*(a) \cdot \pi_Y^*(b)$ where

$a \in h^*(X,A), \quad b \in h^*(Y,B)$

$\pi_X : (X\times Y,\ A\times Y) \longrightarrow (X,A)$

$\pi_Y : (X\times Y,\ X\times B) \longrightarrow (Y,B)$

are the projections and $\cdot$ denotes the multiplication

$$\begin{array}{ccc} h^*(X\times Y,\ A\times Y) \underset{R}{\otimes} h^*(X\times Y,\ X\times B) & \longrightarrow & h^*((X,A) \times (Y,B)) \\ & & \| \\ & & h^*(X\times Y,\ X\times B \cup A\times Y) \end{array}$$

3.7 Note that $h^*(X)$, $h^*(A)$ and $h^*(X,A)$ are all modules over $h^*(X)$. We assume in addition all the homomorphisms in the exact triangle

$$\begin{array}{ccc} h^*(X) & \longrightarrow & h^*(A) \\ & \nwarrow \quad \swarrow \delta & \\ & h^*(X,A) & \end{array}$$

are homomorphisms of $h^*(X)$ modules. In particular, $\delta(x\cdot a) = x\cdot\delta(a)$ for $x \in h^*(X)$, $a \in h^*(A)$.

3.8 Let V and W be two complex G vector bundles over $X \in LC(G)$. Using cross product, we can define a multiplication

$$h^*(V) \underset{R}{\otimes} h^*(W) \xrightarrow{\ \otimes\ } h^*(V\oplus W)$$

denoted by $\otimes$ and written $x \in h^*(V)$, $y \in h^*(W)$ $x \otimes y \in h^*(V\oplus W)$.

The definition of $\otimes$ is this composition:

$$h^*(V) \underset{R}{\otimes} h^*(W) \xrightarrow{\ \times\ } h^*(V\times W) \xrightarrow{\ \tilde{\Delta}^*\ } h^*(V\oplus W)$$

where $\Delta : X \longrightarrow X\times X$ is the diagonal map $\Delta(a) = (a,a)$ and

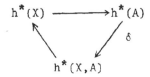

$\tilde{\Delta} : \Delta^*(V\times W) \longrightarrow V\times W$ is the induced map of vector bundles. Note that

$$\Delta^*(V\times W) = V\oplus W.$$

We have distinguished classes $\lambda_V \in h^*(V)$ and $\lambda_W \in h^*(W)$ and <u>assume</u> <u>that</u>

3.9 $$\lambda_V\times\lambda_W = \lambda_{V\times W}; \quad \text{so} \quad \lambda_V\otimes\lambda_W = \lambda_{V\oplus W}.$$

Note that if $s_V : X\longrightarrow V$ and $s_W : X\longrightarrow W$ are the zero sections then

$$(s_V)^*\lambda_V = \lambda_{-1}(V); \quad (s_W)^*\lambda_W = \lambda_{-1}(W)$$

so

$$\lambda_{-1}(V\oplus W) = \lambda_{-1}(V)\cdot\lambda_{-1}(W) \quad \text{by } 3.5\ (*)$$

Thus 1.37 is actually a consequence of 3.9.

3.10 Because of these assumptions in cohomology $h^*(\ )$, we obtain a cross product in homology $h_*(\ )$, i.e., a homomorphism

$$h_*(X,A)\otimes h_*(Y,B) \xrightarrow{\ \times\ } h_*((X,A)\times(Y,B))$$

Of course, A, B, X, and Y are smooth G manifolds. The homomorphism is defined by

$$h^*(T(\dot{X}-\dot{A}))\otimes h^*(T(\dot{Y}-\dot{B})) \xrightarrow{\ \times\ } h^*(T(\dot{X}-\dot{A})\times T(\dot{Y}-\dot{B}))$$

3.11 Having discussed the multiplicative properties of h^* and h_* we can introduce the cap product $\cap$

$$\cap : h_*(X,A) \otimes h^*(X,A) \longrightarrow h_*(X).$$

Let N denote a tubular neighborhood of A in X and define

3.12 $$\Lambda : (T\dot{X})^+ \longrightarrow \{T(\dot{X} - D(\dot{N})) \times (X - D(N)\}^+$$

by

3.13 $$\Lambda(v_x) = (v_x, x) \text{ if } x \notin D(N)$$
$$\Lambda(y) = + \text{ otherwise.}$$

Here $v_x \in T\dot{X}$ is a vector tangent to $\dot{X}$ at $x \in X$. The cap product is then defined by

3.14 $$h^*(T(\dot{X}-D(\dot{N}))) \otimes h^*(X-D(N)) \xrightarrow{\times} h^*(T(\dot{X}-D(\dot{N})) \times (X-D(N)))$$
$$\xrightarrow{\Lambda^*} h^*(T\dot{X}).$$

As a special case observe that if X is a closed manifold, the cap product $h_*(X) \otimes h^*(X) \longrightarrow h_*(X)$ is just

3.15 $$h^*(TX) \otimes_R h^*(X) \xrightarrow{\cdot} h^*(TX)$$

the homomorphism which makes $h^*(TX)$ a module over $h^*(X)$. In this particular case we have Λ defined from TX to $TX \times X$.

To obtain a feel for the cap product, let's examine the case $(X,A) = (D(V),\ S(V))$ where V is a complex G module. Then $h_*(D(V),\ S(V)) = h^*(T(D(\dot{V}) - \dot{Y}))$ where $Y = \{v \in D(V) \mid \|v\| \geq {}^1/_2\}$; so $D(\dot{V}) - \dot{Y} = \{v \in D(V) \mid \|v\| \leq {}^1/_2\}$ and this is G homeomorphic to $D(V)$ so

3.16
$$h_*(D(V),S(V)) = h^*(TD(V))$$
$$h^*(D(V),S(V)) = h^*(V)$$

Now $TD(V) = D(V) \times iV$ where points of iV correspond to vectors tangent to points of $D(V)$.

Since $D(\dot{V})$ is G homeomorphic to V, we identify $TD(\dot{V})$ with $V \times iV$. Let $\rho : TD(V) = D(V) \times iV \longrightarrow iV$ denote the projection. Then we have a commutative diagram

3.17
$$\begin{array}{ccc} TD(\dot{V})^+ & \xrightarrow{\ \Lambda\ } & (TD(V)\times V)^+ \\ \| & & \big\downarrow (\rho\times 1_V)^+ \\ (V\times iV)^+ & \cong & (iV\times V)^+ \end{array}$$

Note that ρ is a G homotopy equivalence. From 3.16 we get this commutative diagram:

$$\begin{array}{ccccc} h_*(D(V),\ S(V)) \otimes h^*(D(V),\ S(V)) & & \xrightarrow{\ \cap\ } & & h_*(D(V)) \\ \| & & & & \| \\ h^*(TD(V)) \otimes h^*(V) & \xrightarrow{\ \times\ } & h^*(TD(V)\times V) & \xrightarrow{\ \Lambda^*\ } & h_*(TD(\dot{V})) \\ \| & & & & \| \\ h^*(iV) \otimes h^*(V) & & \xrightarrow{\ \times\ } & & h^*(iV\times V) \end{array}$$

The first square is commutative by definition, the second by 3.17.

Corollary 3.18. *Let* V *be a complex* S^1 *module. We may identify* $h_*(D(V), S(V))$ *with* $h^*(iV)$ *and* $h^*(D(V), S(V))$ *with* $h^*(V)$ *in such a way that cap product corresponds to cross product. In particular cap product is an isomorphism.*

Proof: The first statement is a consequence of 3.17. The second follows from the first because $\times$ is an isomorphism. This is a consequence of the fact that $h^*(iV), h^*(V)$ and $h^*(iV \times V)$ are all isomorphic to R with generators λ_{iV}, λ_V and $\lambda_{iV \oplus V} = \lambda_{iV} \times \lambda_V$ (by 3.9).

4. Fixed point free actions.

In this and subsequent sections of Part I, we take $h^* = K^*_{S^1}(\ )_p$.

We define an index $Id^X_{F/R}: h_1(X) \longrightarrow F/R$ defined only on fixed point free manifolds and study some homological properties of F/R. We then prove that the homomorphism $h_*(X,\partial X) \longrightarrow Hom_R(h^*(X,\partial X), F/R)$ induced by cap product and $Id^X_{F/R}$ is an isomorphism.

Proposition 4.1. *If M is a complex S^1 module with non-zero Euler class $\lambda_{-1}(M) \neq 0$ then the homology and cohomology long exact sequences for the pair (DM, SM) reduce to*

(4.2) $$0 \longrightarrow R \xrightarrow{\cdot\lambda_{-1}(\bar{M})} R \longrightarrow h_1(SM) \longrightarrow 0$$

(4.3) $$0 \longrightarrow R \xrightarrow{\cdot\lambda_{-1}(M)} R \xrightarrow{\pi^*} h^0(SM) \longrightarrow 0$$

where $\bar{M}$ denotes the complex conjugate of M and $\pi: SM \longrightarrow$ point.

Proof: We prove the Proposition for homology. From the long exact sequence we get the exact sequence

$$0 \longrightarrow h_0(SM) \longrightarrow h^0(M \oplus \bar{M}) \xrightarrow{j^*} h^0(M) \longrightarrow h_1(SM) \longrightarrow 0$$

where $j: M \longrightarrow M \oplus \bar{M}$ is $j(v) = (v,0)$. We have identified $T\dot{D}(M) = M \otimes C = M \oplus \bar{M}$. Applying the Thom isomorphism gives

$$0 \longrightarrow h_0(SM) \longrightarrow R \xrightarrow{j^*\psi^{\bar{M}}} R \longrightarrow h_1(SM) \longrightarrow 0$$

where $j : \text{point} \to \bar{M}$ is the zero section. $j^*\psi^{\bar{M}}$ is just multiplication by the nonzero class $\lambda_{-1}(\bar{M})$. Since R is an integral domain the result follows.

Suppose $M \subset N$ are complex S^1 modules and $N^{S^1} = 0$ i.e. $\lambda_{-1}(M), \lambda_{-1}(N) \neq 0$. Applying (4.2) to the inclusion $i:(DM, SM) \subset (DN, SN)$ gives a commutative diagram of exact sequences

(4.4)
$$\begin{array}{ccccccccc} 0 & \longrightarrow & R & \xrightarrow{\cdot\lambda_{-1}(\bar{M})} & R & \longrightarrow & h_1(SM) & \longrightarrow & 0 \\ & & \downarrow{\scriptstyle 1} & & \downarrow{\scriptstyle \cdot\frac{\lambda_{-1}(\bar{N})}{\lambda_{-1}(\bar{M})}} & & \downarrow{\scriptstyle i_*} & & \\ 0 & \longrightarrow & R & \xrightarrow{\cdot\lambda_{-1}(\bar{N})} & R & \longrightarrow & h_1(SN) & \longrightarrow & 0 \end{array}$$

Let Λ denote all S^1 modules, with nonzero Euler class, ordered by inclusion. The nonzero Euler classes form a multiplicative closed set in R; so let F denote R localized at this set. If we take the direct limit of $h_0(DN)$, $h_0(DN, SN)$, $h_1(SN)$ over Λ then it follows from (4.4) that the limits are R, F, F/R respectively. We have a commutative diagram of limiting maps

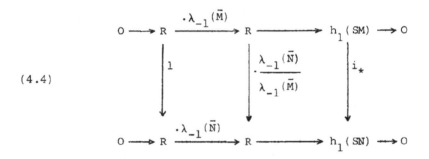

$$\begin{array}{ccccccccc}
0 & \longrightarrow & h_0(DM) & \xrightarrow{j^*} & h_0(DM,SM) & \xrightarrow{\partial} & h_1(SM) & \longrightarrow & 0 \\
 & & \big\downarrow \mathrm{Id}_R & & \big\downarrow \mathrm{Id}_F & & \big\downarrow \mathrm{Id}_{F/R} & & \\
0 & \longrightarrow & R & \longrightarrow & F & \longrightarrow & F/R & \longrightarrow & 0
\end{array} \qquad (4.5)$$

where Id_R is the Atiyah-Singer Index homomorphism (tensored with Q). Identifying $h_0(DM) = h_0(DM,SM) = R$ as in Proposition 4.1 gives

$$\begin{array}{ccccccccc}
0 & \longrightarrow & R & \xrightarrow{\lambda_{-1}(\bar{M})} & R & \longrightarrow & h_1(SM) & \longrightarrow & 0 \\
 & & \big\downarrow 1 & & \big\downarrow \frac{1}{\lambda_{-1}(\bar{M})} & & \big\downarrow \mathrm{Id}_{F/R} & & \\
0 & \longrightarrow & R & \xrightarrow{i} & F & \longrightarrow & F/R & \longrightarrow & 0
\end{array} \qquad (4.6)$$

where $h_1(SM)$, $\mathrm{Id}_{F/R}$ are completely determined by the diagram.

Proposition 4.7. *$\mathrm{Id}_{F/R}$ is injective and for any $h^0(SM)$ module A*

$$(\mathrm{Id}_{F/R})_* : \mathrm{Hom}_R(A, h_1(SM)) \longrightarrow \mathrm{Hom}_R(A, F/R)$$

is an isomorphism.

Note: A is an R module via the projection $\pi^* : R \to h^0(SM)$.

Proof: Injectivity follows from (4.6), thus $(\mathrm{Id}_{F/R})_*$ is also injective. Note that $h_1(SM) = R/(\lambda_{-1}(\bar{M})) = R/(\lambda_{-1}(M))$

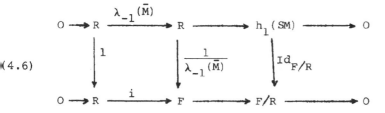

since $\lambda_{-1}(M)$, $\lambda_{-1}(\bar{M})$ differ by a unit. Let $f: A \longrightarrow F/R$ be a R homomorphism. For $a \in A$ let $f(a) = r/s + R$ for some $r, s \in R$. Since $\lambda_{-1}(\bar{M})a = 0$, we have that $\lambda_{-1}(\bar{M})r/s \in R$. But

the image under $Id_{F/R}$ of $\lambda_{-1}(\bar{M})r/s + (\lambda_{-1}(\bar{M})) \in R/(\lambda_{-1}(\bar{M})) = h_1(SM)$ is $r/s + R$ so f takes values in $h_1(SM)$. This means that $(Id_{F/R})_*$ is onto.

<u>Proposition</u> 4.8. <u>$Hom_R(\ ,F)$ and $Hom_R(\ ,F/R)$ are exact functors on the class of R modules with the property that</u> $A \otimes_R F = 0$.

<u>Proof</u>: Let $\bar{F}$ denote the field of fractions of R, then $R \subset F \subset \bar{F}$. Assume $A \otimes_R F = 0$ and consider a R homomorphism $f: A \longrightarrow \bar{F}/F$. If $a \in A$ there is $\lambda \in R$, invertible in F, such that $\lambda a = 0$. Thus $\lambda f(a) = F$, but then $f(a) \cdot = F$ so f is the zero morphism. This means that $Hom_R(A, \bar{F}/F) = 0$. From the exact sequence $0 \longrightarrow F \longrightarrow \bar{F} \longrightarrow \bar{F}/F \longrightarrow 0$ we get an exact sequence $0 \longrightarrow Ext^1_R(A,F) \longrightarrow Ext^1_R(A,\bar{F})$, but since R is a p.i.d. $\bar{F}$ is injective, so $Ext^1_R(A,F) = 0$. It now follows from $0 \longrightarrow R \longrightarrow F \longrightarrow F/R \longrightarrow 0$ that $Ext^1_R(A,F/R) = 0$.

Let X be a compact smooth fixed point free S^1 manifold. Embed X in a complex representation $M \oplus C^n$ where $M^{S^1} = 0$. Projection onto M induces a S^1 map $X \longrightarrow M$. Note that the image of this map does not contain $0 \in M$; so by pushing vectors we get an S^1 map

(4.9) $$I:X \longrightarrow S(M).$$

I induces a homomorphism

$$I_*:h_1(X) \longrightarrow h_1(SM).$$

We define the torsion index $Id^X_{F/R}$ as the composition $Id_{F/R} \circ I_*$

(4.10) $$Id^X_{F/R} : h_1(X) \longrightarrow F/R.$$

It is an easy exercise to show that $Id^X_{F/R}$ is independent of I.

If we compose cap product with the torsion index

$$h_p(X,\partial X) \otimes_R h^{p+1}(X,\partial X) \longrightarrow h_1(X) \longrightarrow F/R$$

we get an induced homomorphism

(4.11) $$\phi^{(X,\partial X)}_{F/R} : h_*(X,\partial X) \longrightarrow \mathrm{Hom}_R(h^*(X,\partial X), F/R).$$

We shall prove that $\phi^{(X,\partial X)}_{F/R}$ is an isomorphism.

Consider first a simple situation. Suppose Γ is a finite subgroup of S^1 and V a real Γ module. The composition of cross product with the Thom isomorphism

$$K^p_\Gamma(V) \otimes K^p_\Gamma(V) \xrightarrow{\times} K^0_\Gamma \ V\otimes C) \xrightarrow{\psi^{-1}} R(\Gamma)$$

induces the homomorphism

(4.12) $$\omega_{\Gamma}^{V} : K_{\Gamma}^{p}(V) \longrightarrow \mathrm{Hom}_{R(\Gamma)}(K_{\Gamma}^{p}(V), R(\Gamma)).$$

We wish to show that ω_{Γ}^{V} is an isomorphism. Consider first the case when $\Gamma = Z_2$ and $V = R_-$ the real one dimensional Z_2 module on which Z_2 acts by multiplication by -1.

We need these basic facts

(i) $R(Z_2) = Z[t]/(1-t^2)$.

(ii) If X is a compact G space on which G acts freely then $K_G^*(X) = K^*(X/G)$, [4].

Observe that the representation ring $R(1)$ of the trivial group 1 is just the integers. The homomorphism of $R(Z_2)$ modules $\varepsilon : R(Z_2) \longrightarrow Z$ defined by $\varepsilon(t) = 1$ is called the augmentation homomorphism.

Lemma 4.13. *$K_{Z_2}^{0}(R_-)$ is isomorphic to the ideal $(1-t) \subset R(Z_2)$ and $K_{Z_2}^{1}(R_-) = 0$.*

Proof: Since Z_2 acts freely on $S(R_-)$ it follows that $K_{Z_2}^{*}(S(R_-)) = K^*(S(R_-)/Z_2) = K^*(p)$, and $K^0(p) = Z$, $K^1(p) = 0$. Moreover $K_{Z_2}^{*}(D(R_-)) = K_{Z_2}^{*}(p)$ so $K_{Z_2}^{0}(D(R_-)) = R(Z_2)$, $K_{Z_2}^{1}(D(R_-)) = 0$ and the restriction homomorphism $j^* : K_{Z_2}^{0}(D(R_-)) \longrightarrow K_{Z_2}^{0}(S(R_-))$ is simply ε. Since ε is an epimorphism the long exact sequence for the pair $(D(R_-), S(R_-))$ is reduced to

$$0 \longrightarrow K^0_{Z_2}(R_-) \xrightarrow{s^*} R(Z_2) \xrightarrow{\epsilon} Z \longrightarrow 0;$$

and $K^1_{Z_2}(R_-) = 0$; $s: p \longrightarrow R_-$ denotes the zero section. But then $K^0_{Z_2}(R_-)$ is the kernel of ϵ which is the ideal $(1-t)$.

By the naturality of cross product we have a commutative diagram

$$\begin{array}{ccccc}
K^*_{Z_2}(R_-) \otimes_{R(Z_2)} K^*_{Z_2}(R_-) & \xrightarrow{\times} & K^*_{Z_2}(R^2_-) & \xrightarrow{\psi^{-1}} & R(Z_2) \\
\downarrow{\scriptstyle s^* \otimes s^*} & & \downarrow{\scriptstyle s^*} & \swarrow{\scriptstyle \cdot(1-t)} & \\
R(Z_2) \otimes_{R(Z_2)} R(Z_2) & \longrightarrow & R(Z_2) & &
\end{array}$$

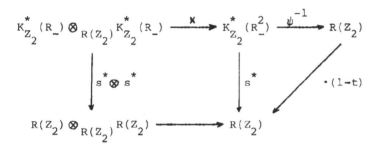

where ψ is the Thom isomorphism for $R^2_- = M_1$ and $1-t = \lambda_{-1}(R^2_-)$. Let g denote the generator in $K^*_{Z_2}(R_-)$ with $s^*(g) = 1-t$, since $R(Z_2) = Z[t]/(1-t^2)$ it follows from 4.13 that $(1+t)\cdot g = 0$ so $(1+t)\cdot\psi^{-1}(g \times g) = 0$. Thus $\psi^{-1}(g \times g) = (1-t)\mu$, for some $\mu \in R(Z_2)$. Moreover we have that $(1-t)^2\mu = (1-t)\psi^{-1}(g \times g) = (s^*(g))^2 = (1-t)^2$, this means $2(1-t)\mu = 2(1-t)$ i.e. $(1-t)\mu = 1-t$. But then $\psi^{-1}(g \times g) = 1-t$. With this in hand we show

Lemma 4.14. $\omega_{Z_2}^{R_-} : K_{Z_2}^{*}(R_-) \longrightarrow \mathrm{Hom}_{R(Z_2)}(K_{Z_2}^{*}(R_-), R(Z_2))$ *is an isomorphism.*

Proof: Let $\lambda g \in K_{Z_2}^{*}(R_-)$, $\lambda \in R(Z_2)$ then

$$\omega_{Z_2}^{R_-}(\lambda g)[g] = \lambda\omega_{Z_2}^{R_-}(g)[g] = \lambda\psi^{-1}(g \times g) = \lambda(1-t).$$

Since $K_{Z_2}^{0}(R_-) \subset R(Z_2)$, by identifying g with $1-t$ it follows that $\omega_{Z_2}^{R_-}(\lambda g)[g] = \lambda g$, thus $\omega_{Z_2}^{R_-}(\lambda g) = 0$ iff $\lambda g = 0$; so $\omega_{Z_2}^{R_-}$ is injective. If $f:K_{Z_2}^{*}(R_-) \longrightarrow R(Z_2)$ is a homomorphism then $(1+t)f(g) = 0$ since $(1+t)\cdot(g)=0$; so $f(g) = \lambda(1-t)$ where $\lambda \in R(Z_2)$. But then $\omega_{Z_2}^{R_-}(\lambda g) = f$; hence, $\omega_{Z_2}^{R_-}$ is an epimorphism.

Let Γ be a finite cyclic group of even order, then there is a surjection $d:\Gamma \longrightarrow Z_2$. For notational convenience we denote by R_- the real Γ module $d^{*}R_-$. By [10] we have $K_{\Gamma}^{*}(R_-) = K_{Z_2}^{*}(R_-) \otimes_{R(Z_2)} R(\Gamma)$.

Corollary 4.15. *If* Γ *has even order, then* $\omega_{\Gamma}^{R_-}$ *is an isomorphism.*

Proof: We need only note that $\omega_{\Gamma}^{R-} = \omega_{Z_2}^{R-} \otimes 1_{R(\Gamma)}$.

Lemma 4.16. If Γ is a finite cyclic group and V is a real Γ module then the homomorphism

$$K_V^*(V) \otimes_{R(\Gamma)} K_\Gamma^*(V) \xrightarrow{\times} K_\Gamma^*(V \otimes C) \xrightarrow{\psi^{-1}} R(\Gamma)$$

induces an isomorphism

$$\omega_\Gamma^V : K_\Gamma^*(V) \longrightarrow \mathrm{Hom}_{R(\Gamma)}(K_\Gamma^*(V), R(\Gamma)).$$

Proof: If Γ is of even order, then V is of the form V_1, $V_1 \oplus R$, $V_1 \oplus R_-$, or $V_1 \oplus R \oplus R_-$ where V_1 admits the structure of a complex Γ module; so by the Thom isomorphism and suspension $\psi^{-1} \circ \times$ is equivalent to multiplication in $K_\Gamma^0(0)$

$$K_\Gamma^0(0) \otimes_{R(\Gamma)} K_\Gamma^0(0) \xrightarrow{\cdot} K_\Gamma^0(0)$$

in the first two cases and

$$K_\Gamma^0(R_-) \otimes_{R(\Gamma)} K_\Gamma^0(R_-) \xrightarrow{\times} K_\Gamma(R_-^2) \xrightarrow{\psi^{-1}} R(\Gamma)$$

in the last two. It is clear that ω_Γ^0 is an isomorphism, ω_Γ^{R-} is an isomorphism by Corollary 4.15. If Γ has odd order then the situation is more simple since V is then of the form V_1 or

$V_1 \oplus R$. Thus by suspension and the Thom isomorphism ω_Γ^V is equivalent to ω_Γ^0.

We have the isomorphism

$$\begin{array}{ccccc}
K_{p+1}^{S^1}(S^1 \times_\Gamma (DV,SV)) \otimes_{R(S^1)} K_{S^1}^p(S^1 \times_\Gamma (DV,SV)) & \xrightarrow{\cap} & K_1^{S^1}(S^1 \times_\Gamma DV) & \xrightarrow{\pi_*} & K_1^{S^1}(S^1/\Gamma) \\
\cong & & \cong & & \cong \\
K_\Gamma^p(V) \otimes_{R(\Gamma)} K_\Gamma^p(V) & \xrightarrow{\times} & K_\Gamma^0(V \otimes C) & \xrightarrow{\psi^{-1}} & R(\Gamma)
\end{array}$$

where $\pi: S^1 \times_\Gamma D(V) \longrightarrow S^1/\Gamma$ is the projection. This follows from the following identification

$$K_{p+1}^{S^1}(S^1 \times_\Gamma (DV,SV)) = K_{S^1}^{p+1}(R \times S^1 \times_\Gamma V) = K_{S^1}^p(S^1 \times_\Gamma V) = K_\Gamma^p(V).$$

Note the general equality $K_G^*(G \times_\Gamma X) = K_\Gamma^*(X)$. See [10] for this equality.

Similarly $K_{S^1}^p(S^1 \times_\Gamma (DV,SV)) = K_\Gamma^p(V)$, $K_1^{S^1}(S^1 \times_\Gamma DV) = K_\Gamma^o(V \otimes C)$, $K_1^{S^1}(S^1/\Gamma) = R(\Gamma)$. With these identifications cap product becomes cross product.

Thus Lemma 4.16 can be interpreted as saying that

$$K_*^{S^1}(S^1 \times_\Gamma (DV,SV)) \longrightarrow \mathrm{Hom}_{R(S^1)}(K_{S^1}^*(S^1 \times_\Gamma (DV,SV), K_1^{S^1}(S^1/\Gamma))$$

is an isomorphism.

Note that if the order of Γ is n then $S^1/\Gamma = S(M_n)$ so changing to the theory h^*, taking $h_1(S(M_n))$ to the limit and applying Proposition 4.7 we have

Theorem 4.17. *If Γ is a discrete subgroup of S^1 and V is a real Γ module then*

$$\phi_{F/R}: h_*(S^1 \times_\Gamma (DV, SV)) \longrightarrow \mathrm{Hom}_R(h^{*+1}(S^1 \times_\Gamma (DV, SV)), F/R)$$

is an isomorphism.

Suppose $X \in D_C(S^1)$ and Z is obtained from X by attaching an S^1 handle H to X. Suppose that the isotropy group associated to the handle H is $\Gamma \subset S^1$, $\Gamma \neq S^1$. Then there are two real Γ modules V and W such that if

$$\text{(i)} \quad H = \{(v,w) \in V \oplus W \mid \|v\| \le 1, \|w\| \le 1\}$$

$$\text{(ii)} \quad H_0 = \{(v,w) \in V \oplus W \mid \|v\| = 1, \|w\| \le 1\}$$

then $H = S^1 \times_\Gamma H$ and $H \cap X = S^1 \times_\Gamma H_0$.

Let $i: S^1 \times_\Gamma (DV, SV) \longrightarrow (H, H \cap X)$ be the inclusion. Then i is a S^1 homotopy equivalence so it induces isomorphisms in homology and cohomology. Also the inclusion $i: (H, H \cap X) \longrightarrow (Z, X)$ induces isomorphisms in homology and cohomology by excision. Thus Theorem 4.17 gives

Corollary 4.18. *Let Z be obtained from X by attaching an S^1 handle with associated isotropy group* $\Gamma \subset S^1$, $\Gamma \neq S^1$, _then_

$$\phi_{F/R}^{(Z,X)} : h_*(Z,X) \longrightarrow \mathrm{Hom}_R(h^{*+1}(Z,X), F/R)$$

is an isomorphism.

Before we globalize Corollary 4.18 we need a version of the Atiyah-Segal Localization Theorem.

Lemma 4.19. _Let_ X _be a_ S^1 _space which can be embedded equivariantly in_ $M \oplus C^n$, M _a complex_ S^1 _module with nonzero Euler class. If_ Y _is a closed invariant subspace of_ X _containing the fixed point set then_ $h^*(X,Y)$ _is annihilated by_ $\lambda_{-1}(M)$. _Moreover if, in addition,_ (X,Y) _is a smooth pair then_ $h_*(X,Y)$ _is annihilated by_ $\lambda_{-1}(M)$.

Proof: Composing the embedding with projection onto M induces a S^1 map $X \longrightarrow M$. Since $X-Y$ has no fixed points its image misses $0 \in M$ so by pushing vectors onto $S(M)$ we get a S^1 map $X-Y \longrightarrow S(M)$ making $h^*(X,Y)$ into a $h^0(SM)$ module. The R module structure of $h^*(X,Y)$ may be recovered via $\pi^*: R \longrightarrow h^0(SM)$. But $\lambda_{-1}(M)$ annihilates the unitary ring $h^0(SM)$ so it annihilates $h^*(X,Y)$. The argument for a smooth pair is similar.

We may now apply Proposition 4.8 to the homology an cohomology of handles. Suppose that X is a compact S^1 manifold with no fixed points. There is a decreasing filtration

$X = X_0 \supset X_1 \supset \dots \supset X_n = \partial X$ such that X_i is obtained from X_{i+1} by attaching a single S^1 handle H_i of orbit type $S(M_{n_i})$, $n_i \neq 0$, i.e. the isotropy group associated to H_i is cyclic of order n_i for some n_i depending upon i. We have two exact triangles

(4.20)
$$\begin{array}{ccc} h_*(X,X_{i+1}) & \longrightarrow & h_*(X,X_i) \\ & \nwarrow \quad \swarrow & \\ & h_*(X_i,X_{i+1}) & \end{array}$$

(4.21)
$$\begin{array}{ccc} \mathrm{Hom}_R(h^*(X,X_{i+1}),F/R) & \text{——} & \mathrm{Hom}_R(h^*(X,X_i),F/R) \\ & \nwarrow \quad \swarrow & \\ & \mathrm{Hom}_R(h^*(X_i,X_{i+1}),F/R) & \end{array}$$

and a morphism $\phi_{F/R}$ from (4.20) to (4.21). $\phi_{F/R}^{(X_i,X_{i+1})}$ is an isomorphism by Corollary 4.18. $\phi_{F/R}^{(X,X_1)}$ is an isomorphism by the same reason so by induction we have

<u>Theorem</u> 4.22. <u>If</u> X <u>is a compact fixed point free</u> S^1 <u>manifold then</u>

$$\phi_{F/R}^{(X,\partial X)} : h_*(X,\partial X) \longrightarrow \mathrm{Hom}_R(h^{*+1}(X,\partial X),F/R)$$

<u>is an isomorphism</u>.

5. The Universal coefficient theorem

We prove here the exact sequence

$$0 \longrightarrow \mathrm{Ext}^1_R(h^{q+1}(X),R) \longrightarrow h_q(X) \xrightarrow{\phi^X} \mathrm{Hom}_R(h^q(X),R) \longrightarrow 0$$

where $\phi^X(a)(b) = \mathrm{Id}^X_R(a \cap b)$.

Assume first that S^1 acts trivially on X. Then

$$K^*_{S^1}(X) = K^*(X) \otimes R(S^1) \qquad [4].$$

Since the ring $R = R(S^1) \otimes Q$ contains the rationals, it follows that

$$h^*(X) = [K^*(X) \otimes Q] \otimes_Q R$$

is a free R module. Define $\tilde{K}^*(X) = K^*(X) \otimes Q$. Let

$$\phi^X_K : \tilde{K}^*(TX) \longrightarrow \mathrm{Hom}_Q(\tilde{K}^*(X),Q)$$

denote the homomorphism defined by

$$\phi^X_K(x)[y] = \mathrm{Id}^X \otimes 1_Q(x \cdot y)$$

where $x \in \tilde{K}^*(TX)$, $y \in \tilde{K}^*(X)$ and $\mathrm{Id}^X : K^*(TX) \longrightarrow K^*(p) = Z$ is the Atiyah-Singer homomorphism. In this case $\mathrm{Id}^X_R = \mathrm{Id}^X \otimes_{\mathbb{Z}} 1_R$; so $\phi^X = \phi^X_K \otimes 1_R$. See [3], p. 538.

<u>Theorem</u> 5.1. ϕ^X <u>is an isomorphism</u>.

<u>Proof</u>: We only need to show that ϕ^X_K is an isomorphism. Suppose the dimension of X is n. If $\alpha \in H^*(X)$ and $[X] \in H_n(X)=$ denotes the orientation class of X, set

$\alpha[X] = \alpha_n \cap [X] \in H_0(X) = Z$ where α_n is the component of α in dimension n. Let $\psi: H^*(X,Q) \longrightarrow H^*(TX,Q)$ denote the Thom isomorphism for rational cohomology, $ch: K^*(X) \otimes Q \longrightarrow H^*(TX,Q)$ the Chern character isomorphism and $I(X) \in H^*(X,Q)$ the cohomology class introduced in [5]. Concerning the class $I(X)$, we only need that it is a unit of $H^*(X,Q)$ and denote its inverse by $I(X)^{-1}$.

In terms of this information, an explicit formula for the homomorphism Id^X is given in [5] page 55. If $\mu \in \tilde{K}^*(TX)$

$$Id^X(\mu) = (-1)^{n(n+1)/2} \left\{ \psi^{-1} ch(\mu) \cdot I(X) \right\} [X].$$

Let $\phi_H^X: H^*(X,Q) \longrightarrow Hom_Q(H^*(X,Q),Q)$ denote the homomorphism defined by

$$\phi_H^X(x)[y] = x \cdot y[X].$$

It is a well known consequence of Poincaré Duality and the universal coefficient theorem, that ϕ_H^X is an isomorphism.

Define $T: \tilde{K}^*(TX) \longrightarrow H^*(X,Q)$ by $T = \psi^{-1} ch$. Define $S: Hom_Q(\tilde{K}^*(X),Q) \longrightarrow Hom_Q(H^*(X,Q),Q)$ by $S(f)(x) = f ch^{-1}(x \cdot I(X)^{-1})$. Then we have a commutative diagram

$$\begin{array}{ccc}
\tilde{K}^*(TX) & \xrightarrow{\phi_K^X} & Hom_Q(\tilde{K}^*(X),Q) \\
\Big\downarrow T & & \Big\downarrow S \\
H^*(X,Q) & \xrightarrow{\phi_H^X} & Hom_Q(H^*(X,Q),Q)
\end{array}$$

Since T, ϕ_H^X and S are isomorphisms, so is ϕ_K^X.

We have to chase some diagrams now: Let X be a closed S^1 manifold. N denotes a closed tubular neighborhood of X^{S^1}. We have a commutative diagram

$$\begin{array}{ccccc}
h^p(TX)\otimes_R h^q(N) & \xrightarrow{1\otimes\delta} & h^p(TX)\otimes_R h^{q-1}(X-N) & \xrightarrow{i^*\otimes 1} & h^p(T(X-\dot{N}))\otimes_R h^{q-1}(X-N) \\
\downarrow \times & & \downarrow \times & & \downarrow \times \\
h^{p-q}(TX\times N) & \xrightarrow{\delta} & h^{p-q-1}(TX\times(X-N)) & \xrightarrow{i^*} & h^{p-q-1}(T(X-\dot{N})\times(X-N)) \\
\downarrow \Delta^* & & \downarrow \Delta^* & & \downarrow \Delta^* \\
h^{p-q}(TN) & \xrightarrow{\delta} & h^{p-q-1}(T(X-N)) & \xrightarrow{1} & h^{p-q-1}(T(X-N))
\end{array}$$

where Δ is the diagonal map taking a vector v_x over x into (v_x, x). Note that the right vertical sequence is simply cap product for the pair $(X-\dot{N}, \partial N)$. We interpret i^* in the top line as $j_*: h_p(X) \longrightarrow h_p(X,N)$ and δ in the bottom line as $\partial: h_{p-q}(X, X-\dot{N}) \longrightarrow h_{p-q-1}(X-\dot{N})$. Let $e: (X-\dot{N}, \partial N) \subset (X,N)$ denote excision. If $x \in h_p(X)$, $y \in h^q(N)$, then the above diagram translates into

$$\text{(i)} \qquad \partial\Delta^*(x \times y) = (e_*^{-1} j_* x) \cap (e^* \delta y)$$

Consider the following commutative diagram

$$\begin{array}{ccc}
h^p(TX)\otimes_R h^q(X) & \xrightarrow{1\otimes i^*} & h^p(TX)\otimes_R h^q(N) \\
\downarrow{\times} & & \downarrow{\times} \\
h^{p-q}(TX\times X) & \xrightarrow{i^*} & h^{p-q}(TX\times N) \\
\downarrow{\Delta^*} & & \downarrow{\Delta^*} \\
h^{p-q}(TX) & \xrightarrow{i^*} & h^{p-q}(TN)
\end{array}$$

The left vertical sequence is cap product for X. Here we interpret i^* in the bottom line as $j_*: h_{p-q}(X) \longrightarrow h_{p-q}(X, X-\dot{N})$. If $x \in h_p(X)$, $z \in h^q(X)$ then the above diagram gives

$$\text{(ii)} \qquad j_*(x \cap z) = \Delta^*(x \times i^* z).$$

Embed X in $M \oplus C^n$ where M is a complex S^1 module with $M^{S^1} = 0$. Projecting onto M gives a map $I: X \to M$. Note that $0 \in M$ is not in the image of $X-\dot{N}$. By a smooth equivariant deformation we may assume that $I: X \longrightarrow DM$ and $I(X-\dot{N}) \subset SM$; so we have a commutative diagram of exact sequences

$$\begin{array}{ccccccccc}
 & & h_0(X) & \xrightarrow{j_*} & h_0(X, X-\dot{N}) & \xrightarrow{\partial} & h_1(X-\dot{N}) & & \\
 & & \downarrow{I_*} & & \downarrow{I_*} & & \downarrow{I_*} & & \\
0 & \longrightarrow & h_0(DM) & \xrightarrow{j_*} & h_0(DM, SM) & \xrightarrow{\partial} & h_1(SM) & \longrightarrow & 0
\end{array}$$

Composing the above diagram with (4.5) gives

$$\text{(iii)}\qquad \begin{array}{ccccccccc} & & h_0(X) & \xrightarrow{j_*} & h_0(X, X-\dot{N}) & \xrightarrow{\partial} & h_1(X-\dot{N}) & & \\ & & \Big\downarrow \mathrm{Id}_R & & \Big\downarrow \mathrm{Id}_F & & \Big\downarrow \mathrm{Id}_{F/R} & & \\ 0 & \longrightarrow & R & \xrightarrow{i} & F & \xrightarrow{\pi} & F/R & \longrightarrow & 0 \end{array}$$

where Id_R is the homomorphism $\mathrm{Id}^X_{S^1} \otimes_{R(S^1)} 1_R$.

Let $x \in h_p(X)$, $y \in h^p(N)$; by Lemma 4.19 $h^{p+1}(X,N)$ is annihilated by $\lambda = \lambda_{-1}(M)$ so there is $z \in h^p(X)$ with $i^* z = \lambda y$, then

$$\begin{aligned} \pi \frac{1}{\lambda} i \mathrm{Id}_R(x \cap z) &\overset{(iii)}{=} \pi \frac{1}{\lambda} \mathrm{Id}_F j_*(x \cap z) \overset{(ii)}{=} \pi \frac{1}{\lambda} \mathrm{Id}_F \Delta^*(x \times \lambda y) \\ &= \pi \mathrm{Id}_F \Delta^*(x \times y) \overset{(iii)}{=} \mathrm{Id}_{F/R} \partial \Delta^*(x \times y) \\ &\overset{(i)}{=} \mathrm{Id}_{F/R}((e_*^{-1} j_* x) \cap (e^* \delta y)). \end{aligned} \tag{5.2}$$

We define a homomorphism

$$\omega : \mathrm{Hom}_R(h^*(X), R) \longrightarrow \mathrm{Hom}_R(K, F/R)$$

where $K \subset h^*(X,N)$ is the image of δ. Consider the diagram of exact sequences where $\bar{f}$ and $\omega(f)$ are defined by f as follow

$$\text{(5.3)}\qquad \begin{array}{ccccccccc} & & h^*(X) & \xrightarrow{i^*} & h^*(N) & \xrightarrow{\delta} & K & \longrightarrow & 0 \\ & & \Big\downarrow f & & \Big\downarrow \bar{f} & & \Big\downarrow \omega(f) & & \\ 0 & \longrightarrow & R & \xrightarrow{i} & F & \xrightarrow{\pi} & F/R & \longrightarrow & 0 \end{array}$$

Since $\lambda = \lambda_{-1}(M)$ annihilates K there is a unique homomorphism $\bar{f}$ which makes the diagram commutative, $\bar{f}(y) = \frac{1}{\lambda} if(z)$ where $i^* z = \lambda y$. Now $\omega(f)$ is uniquely defined by $\omega(f)(\delta y) = \pi \frac{1}{\lambda} if(z)$. Moreover since $h^*(N) = h^*(X^{S^1})$ is free, it follows that ω is an epimorphism.

Define

$$\phi^X : h_*(X) \longrightarrow \mathrm{Hom}_R(h^*(X), R)$$

by

$$\phi^X(x)[z] = \mathrm{Id}_R(x \cap z)$$

and

$$\phi : h_*(X, X^{S^1}) = h_*(X, N) \longrightarrow \mathrm{Hom}_R(K, F/R)$$

by

$$\phi(x)[\delta y] = \mathrm{Id}_{F/R}(e_*^{-1} x \cap e^* \delta y)$$

then (5.2) becomes

$$\phi j_* = \omega \phi^X \tag{5.4}$$

<u>Proposition</u> 5.5. <u>If</u> X <u>is a closed</u> S^1 <u>manifold then</u>

$$\begin{array}{ccccccccc}
0 & \longrightarrow & h_*(X^{S^1}) & \xrightarrow{\;i_*\;} & h_*(X) & \xrightarrow{\;j_*\;} & h_*(X,X^{S^1}) & \longrightarrow & 0 \\
 & & \downarrow{\scriptstyle \phi^{X^{S^1}}} & & \downarrow{\scriptstyle \phi^{X}} & & \downarrow{\scriptstyle \phi} & & \\
0 & \longrightarrow & \mathrm{Hom}_R(h^*(X^{S^1}),R) & \xrightarrow{\;i^{**}\;} & \mathrm{Hom}_R(h^*(X),R) & \xrightarrow{\;\omega\;} & \mathrm{Hom}_R(K,F/R) & \longrightarrow & 0
\end{array}$$

is a commutative diagram of exact sequences.

Proof. The first square is commutative by the naturality of cap product and the second by (5.4). The top horizontal sequence is exact since $h_*(X^{S^1})$ is free and $h_*(X,X^{S^1})$ is torsion. i^{**} is injective because $\operatorname{cok} i^* = K$ is torsion, also ω was already shown to be onto. It remains to show exactness at $\mathrm{Hom}_R(h^*(X),R)$. If $f:h^*(X^{S^1}) \to R$ is a homomorphism then by (5.3)

$$\omega(i^{**}f)[\delta y] = \pi \frac{1}{\lambda} i f i^*(z) = \pi \frac{1}{\lambda} i f(\lambda y)$$

$$= \pi i f(y) = 0, \text{ where } i^* z = \lambda y.$$

Also if $f:h^*(X) \to R$ and $\omega f(\delta y) = \pi \frac{1}{\lambda} i f(z) = 0$, then from (5.3) it follows that $\bar{f}:h^*(X^{S^1}) \to F$ takes values in R so $i^{**}(\bar{f}) = f$

Theorem 5.6. *If X is a closed S^1 manifold then there is a split exact sequence*

$$0 \longrightarrow \mathrm{Ext}^1_R(h^{q+1}(X),R) \longrightarrow h_q(X) \xrightarrow{\phi^X} \mathrm{Hom}_R(h^q(X),R) \longrightarrow 0.$$

Proof. By Theorem 5.1 $\phi^{X^{S^1}}$ is an isomorphism. Now ϕ is the composition

$$h_*(X,N) \underset{=}{\overset{e_*^{-1}}{=}} h_*(X-\dot{N},\partial N) \xrightarrow{\phi_{F/R}^{(X-\dot{N},\partial N)}} \mathrm{Hom}_R(h^*(X-\dot{N},\partial N),F/R)$$

$$\overset{e^*}{=} \mathrm{Hom}_R(h^*(X,N),F/R) \xrightarrow{q^*} \mathrm{Hom}_R(K,F/R)$$

where $q:K \subset h^*(X,N)$. By Proposition 4.8 q^* is onto and by Theorem 4.22 $\phi_{F/R}^{(X-\dot{N},\partial N)}$ is an isomorphism so ϕ is an epimorphism. It follows from Proposition 5.5 that ϕ^X is an epimorphism.

It remains to prove the injective part. Since $\phi^{X^{S^1}}$ and $\phi_{F/R}^{(X-\dot{N},\partial N)}$ are isomorphisms, the kernel of ϕ^X may be identified with the kernel of q^*. Consider the exact sequence $0 \to K \to h^{q+1}(X,N) \to T \to 0$. The kernel of ϕ^X is isomorphic to $\mathrm{Hom}_R(T,F/R)$. From the exact sequence $0 \to R \to F \to F/R \to 0$, it follows that $\mathrm{Hom}_R(T,F/R) = \mathrm{Ext}^1_R(T,R)$ by Proposition 4.8. Moreover, if we consider the exact sequence $0 \to T \to h^{q+1}(X) \to L \to 0$ then $L \subset h^{q+1}(X^{S^1})$ is free since R is a p.i.d and $h^{q+1}(X^{S^1})$ is free, so $\mathrm{Ext}^1_R(h^{q+1}(X),R) = \mathrm{Ext}^1_R(T,R)$.

Remark. We indicate how to generalize the universal coefficient theorem for the torus while keeping the homological dimension of R low. Let P be the ideal generated by an

irreducible element in $R(T^n) \otimes Q = Q[s_1, \ldots, s_n, s_1^{-1}, \ldots, s_n^{-1}]$.
Define

$$h^*(X) = (K_{T^n}^*(X) \otimes Q)_p$$

and $R = h^*(\text{point}) = (R(T^n) \otimes Q)_p$. Then R will be a p.i.d. R
will still be a p.i.d. when P is a union of a finite number of
ideals as defined above. In the case $n = 1$ we need not
localize $R(S^1) \otimes Q$ since this is already a p.i.d.

The main problem is to generalize Lemma 4.16. The rest
follows identically as in sections 4 and 5. We have to prove
that if H is a proper subgroup of T^n and V is a real H
module then cross product composed with the Thom isomorphism
induces an isomorphism

$$\omega_H^V : h_H^*(V) \longrightarrow \underset{h_H^*(p)}{\mathrm{Hom}}(h_H^*(V), h_H^*(p))$$

where $h_H^* = K_H^*(\quad)_p$. K_H^* is an algebra over $K_{T^n}^*$ given by the
inclusion $H \subset T^n$.

It is easy to show that $h_H(p) = 0$ if the dimension of
$H < n-1$; so we are done in this case. If $\dim H = n-1$ then one
shows that the problem is equivalent to Lemma 4.16.

6. Poincaré duality

Our aim is to establish an isomorphism

$$\Delta_1^X : K^*_{S^1}(X) \longrightarrow K^*_{S^1}(TX)$$

when X is a smooth closed S^1 manifold such that $|X|$ is a spin ^{c} manifold. In order to do this we must discuss some properties of the group spin ^{c} (m).

Let V be a real vector space of dimension $m = 2n$. We suppose V endowed with the standard inner product with respect to an orthonormal base $e_1, e_2, \cdots, e_m$. Let $A(V)$ denote the Clifford algebra of V [2], [8]. For $v \in V \subset A(V)$

6.1) $$v^2 = -\|v\|^2 \cdot 1$$

where $1 \in A(V)$ is the identity.

$A(V)$ is the direct sum $A^+ \oplus A^-$ where A^+ is spanned by the products $e_{i_1} e_{i_2} \cdots e_{i_k}$ with k even and A^- by the products with odd k. The multiplicative subgroup of $A(V)$ generated by elements of the unit sphere $S(V) \subset A(V)$ is denoted by pin (m). The intersection pin $(m) \cap A^+$ is the group spin (m).

The group spin (m) acts in an obvious manner on $A(V) \otimes C$ giving a linear representation of spin (m). This representation is reducible

$$A(V) \otimes C = \Delta_+ \oplus \Delta_-$$

where Δ_+ is the $+$ eigenspace of

$$(i)^n e_1 e_2 \cdots e_m = \tau$$

and Δ_- is the negative eigenspace of τ.

Observe that $\tau^2 = 1$ and τ commutes with elements of A^+ and so with spin(m) and

$$\tau v = -v\tau \quad \text{for} \quad v \in V.$$

Because of this, left multiplication by $v \in V$, denoted by $L(v)$, maps Δ_+ to Δ_- and vice versa. Let

$$\theta : V \times \Delta_{\pm} \longrightarrow V \times \Delta_{\mp}$$

be the map defined by

(6.2) $$\theta(v,\delta) = (v, L(v)\delta), \quad v \in V, \quad \delta \in \Delta_{\pm}.$$

Then θ is elliptic, i.e., for fixed $v \neq 0$ in V the linear map

$$\theta_v : v \times \Delta_{\pm} \longrightarrow v \times \Delta_{\mp}$$

defined by restricting θ is an isomorphism. This follows from the fact that

$$\theta_v \circ \theta_v(v,\delta) = (v, L(v)L(v)\delta) = (v, -||v||^2\delta)$$

because $L(v) \circ L(v) = L(v^2) = -||v||^2 \cdot 1$ by (7.1).

The generator $\varepsilon = -1 \in A(v)$ of the double covering

$$\pi_1 : \mathrm{spin}(m) \longrightarrow SO(m)$$

acts as multiplication by -1 on Δ_+ and Δ_-. This means that the action of spin(m) on these two representation spaces may be extended to the group

$$\mathrm{spin}^c(m) = \mathrm{spin}(m) \times_{Z_2} S^1 .$$

Here $Z_2 \subset \mathrm{spin}(m)$ is the subgroup generated $-1 \in \mathrm{spin}(m)$ and $Z_2 \subset S^1$ is the subgroup generated by -1 S^1. Explicitly if $[g,t]$ denotes an equivalence class in $\mathrm{spin}^c(m)$ determined by $g \in \mathrm{spin}(m)$ and $t \in S^1 \in C$, then

$$[g,t]\delta = t\cdot(g\cdot\delta) \quad \text{for} \quad \delta \in \Delta_{\pm}$$

Of particular importance to us is the commutative diagram

(6.3)

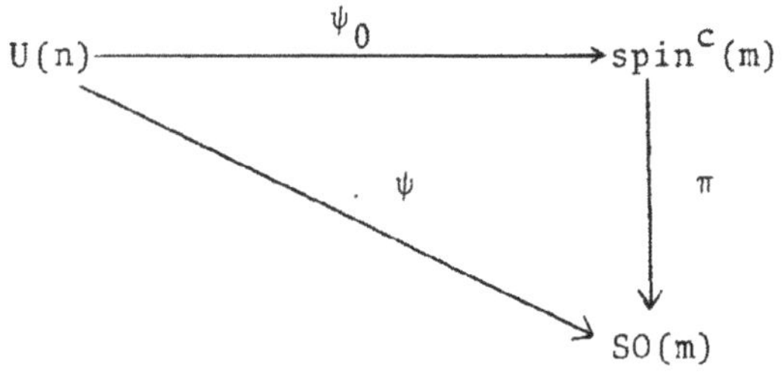

Here $\pi[g,t] = \pi_1(g)$;

$$\psi \operatorname{diag}\left(e^{i\theta_1}, e^{i\theta_2} \cdots e^{i\theta_n}\right)$$

$$= \operatorname{diag}\begin{pmatrix} \cos\theta_j & \sin\theta_j \\ -\sin\theta_j & \cos\theta_j \end{pmatrix} \subset SO(m)$$

$$\psi_0 \operatorname{diag}\left(e^{i\theta_1}, e^{i\theta_2} \cdots e^{i\theta_n}\right)$$

$$= \left[\prod_{j=1}^{n} (\cos\theta_j/2 - \sin\theta_j/2\, e_{2j-1}e_{2j}),\ e^{-i(\Sigma\theta_j/2)}\right]$$

Note that

$$\prod_{j=1}^{n}(\cos\theta_j/2 - \sin\theta_j/2\, e_{2j-1}e_{2j}) \in \text{spin}(m) \subset A(V)$$

so ψ_0 makes sense and $\pi\psi_0 = \psi$.

Observe that $\text{spin}^c(m)$ has a central circle subgroup S^1 and the quotient is $SO(m)$. The orbit map is π.

Moreover, there is an exact sequence of groups

(6.4) $$1 \longrightarrow \text{spin}(m) \xrightarrow{\ i\ } \text{spin}^c(m) \xrightarrow{\ j\ } S^1 \longrightarrow 1$$

$$j[g,t] = t^2$$

and a commutative diagram

(6.5)
$$\begin{array}{ccc} spin^c(m) \times S^1 & \xrightarrow{\quad m_1 \quad} & spin^c(m) \\ \Big\downarrow{\scriptstyle j \times d} & & \Big\downarrow{\scriptstyle j} \\ S^1 \times S^1 & \xrightarrow{\quad m_2 \quad} & S^1 \end{array}$$

Here m_1 is multiplication in $spin^c(m)$, m_2 multiplication in S^1 and d is the squaring map $d(t) = t^2$. Since S^1 is central in $spin^c(m)$, m_1 is a homomorphism of groups.

Let X be a compact G manifold of dimension m. Let W be a (real) G vector bundle over X of dimension k. An orientation for W is a class $\omega_G \in K_G^k(W)$ such that $i^*\omega_G \in K_G^k(W|0)$ generates $K_G^*(W|0)$ freely over $K_G^*(0)$ for every orbit 0. Here i is the inclusion of $W|0$ in W.

<u>Definition: An orientation for X is an orientation</u> $\alpha_G \in K_G^m(TX)$ <u>of the tangent bundle of X.</u>

Observe that if X has a boundary ∂X, then ∂X is oriented by $j^*(\alpha_G)$, $j : \partial X \longrightarrow X$ because

$$K_G^m(TX|_{\partial X}) = K_G^m(T\partial X \times R^1) = K_G^{m-1}(T\partial X).$$

An orientation class α_G provides a Thom homomorphism

$${}^X_1 : K_G^*(X) \longrightarrow K_G^*(TX), \quad \Delta_1^X(\lambda) = \alpha_G . \lambda \ .$$

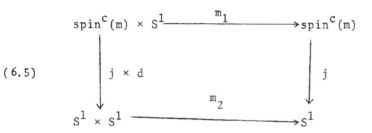

Lemma 6.6 Δ_1^X <u>is an isomorphism</u>.

Proof: Let $\bar{X}$ denote the orbit space of X by G. There are two sheaves over $\bar{X}$, S_q. and T_q whose stalks are respectively

$$S_q(\bar{x}) = K_G^q(Gx)$$

$$T_q(\bar{x}) = K_G^q(TGx)$$

where $\bar{x} \in \bar{X}$ and $Gx \subset X$ is the orbit of $x \in X$ lying over $\bar{x}$.

Multiplication by α_G induces a map of the spectral sequence [7]

$$E_2^{p,q} = H^p(\bar{X}, S_q) \longrightarrow K_G^*(X)$$

to the spectral sequence

$$E_2^{p,q} = H^p(\bar{X}, T_q) \longrightarrow K_G^*(TX)$$

which is an isomorphism on the E_2 level.

Corollary 6.7 $\Delta_1^{\partial X}$ <u>is an isomorphism</u>.

Corollary 6.8 $\Delta_1^{(X,\partial M)} : K_G^*(X, \partial X) \longrightarrow K_G^*(TX, TX|_{\partial X})$ <u>is an isomorphism</u>.

Proof: Multiplication by α_G induces a map of the exact sequence of the pair $(X, \partial X)$ to $(TX, TX|_{\partial X})$ which is an isomorphism on two terms by the preceding. The result follows by the five lemma.

Remark: We could equally well have defined an orientation for X by means of a class $\beta_G \in K_G^*(NX)$ where NX is the normal bundle of X which is equivariantly imbedded in a complex representation space M for G. These are equivalent concepts.

The significance of this remark is that $K_G^*(NX)$ is the equivariant homology of X dual to $K_G^*(X)$ if X is a closed manifold. To see this, note that $X \in M$, $M^+ = S^{2n}$ where M^+ is the one point compactification of M which we assume as complex dimension n. Then by definition

$$(6.9) \qquad K_i^G(X) = K_G^{2n-i}(S^{2n}, S^{2n}-X) = K_G^{2n-i}(NX, NX|_{\partial X})$$

by excision.

The map C which collapses the exterior of the closure $\overline{NX}$ of NX in M^+ induces

$$C^* : K_G^*(\overline{NX}, \partial\overline{NX}) \longrightarrow K_G^*(M^+, +) = R(G).$$

The composition of c^* with the map $n : K_G^*(X) \otimes K_G^*(NX, NX|_{\partial X}) \longrightarrow K_G^*(NX, NX|_{\partial X})$ which exhibits $K_G^*(NX, NX|_{\partial X})$ as a module over $K_G^*(X)$ defines the duality pairing

$$d': K^i_G(X) \otimes K^G_{m-i}(X) \longrightarrow R(G).$$

There is a second duality pairing d which is more appropriately related to our purpose when X is a closed manifold. It is the map

$$d : K^*_G(X) \otimes K^*_G(TX) \longrightarrow R(G)$$

which is defined by

$$d(x \otimes y) = Id^X_G(x \cdot y)$$

for $x \in K^*_G(X)$ and $y \in K^*_G(TX)$.

Remark: When X is a closed manifold, these two pairings are the same. The point is that the pairing d' is more accessible to computation because of the properties of the index homomorphism. To extend the definition of d to spaces X which are not manifolds we must assume that X is imbedded in a complex representation space M of G with an equivariant regular neighborhood $NX \subset M^+$. Then $K^G_i(X) = K^{2n-i}_G(\overline{NX}, \partial\overline{NX})$ and d is defined for such G spaces X as above.

To justify the above remark we offer

Proposition 6.10 If X is a closed G manifold, there is an isomorphism $\phi : K^*_G(TX) \longrightarrow K^*_G(NX)$ which takes d to d'.

Proof: There is a commutative diagram of vector bundles

$$
\begin{array}{ccc}
TNX & \longrightarrow & TX \\
\big\downarrow & & \big\downarrow \pi \\
NX & \longrightarrow & X
\end{array}
$$

Since $NX \subset M$ is an open subset, $TNX = NX \times M$ as a G vector bundle over NX. However, $TNX = \pi^*(NX \otimes C)$. Since TNX is a complex G bundle over NX as well as over TX, we have Thom isomorphisms

$$\psi_1 : K_G^*(NX) \longrightarrow K_G^*(TNX)$$

$$\psi_2 : K_G^*(TX) \longrightarrow K_G^*(TNX)$$

and a commutative diagram

$$
\begin{array}{ccccc}
K_G^*(NX) \otimes K_G^*(X) & \longrightarrow & K_G^*(NX) & \longrightarrow & K_G^*(M) \\
\big\downarrow \psi_1 \otimes 1 & & \big\downarrow \psi_1 & & \big\downarrow \\
K^*(TNX) \otimes K_G^*(X) & \longrightarrow & K_G^*(TNX) & \longrightarrow & K_G^*(TM) \\
\big\downarrow \psi_2 \otimes 1 & & \big\downarrow \psi_2 & & \big\downarrow \\
K_G^*(TX) \otimes K_G^*(X) & \longrightarrow & K_G^*(TX) & \xrightarrow{\ Id_G\ } & R(G)
\end{array}
$$

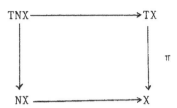

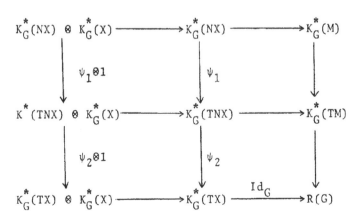

In which all vertical maps are isomorphisms. Since d' is defined by the top row and d by the bottom, the demonstration is complete.

With the equivariant homology of a manifold X defined by 6.9 we obtain Poincare duality free if an orientation is given.

Proposition 6.11 <u>If</u> X <u>is a compact oriented</u> G <u>manifold of dimension</u> m, <u>then</u> X <u>satisfies Poincare duality</u>: $K_i^G(X) = K_G^{m-i}(X,\partial X)$.

Proof: If $\alpha_G \in K_G^m(TX)$ is an orientation then $\Delta_1^X : K_G^*(X,\partial X) \longrightarrow K_G^*(TX,TX|_{\partial X})$ is an isomorphism. But

$$K_G^*(TX,TX|_{\partial X}) = K_G^*(NX,NX|_{\partial X}) = K_*^G(X).$$

In view of Proposition 6.11, we expect the difficulties of studying G actions on manifolds by using Poincare duality to be intimately connected with the existence of an orientation. There is a very general situation in which it is often possible to construct an orientation. This occurs when X is a $spin^c(m)$ manifold. Again, this means there is a principle $spin^c(m)$ bundle P over X such that

$$P \times_{spin^c(m)} V = TX \text{ the tangent bundle of } X.$$

6.12 We assume: that G acts on the left on P, commutes with the right action of $\mathrm{spin}^c(m)$ and the induced G action on $P \times_{\mathrm{spin}^c(m)} V$ agrees with the natural G action on TX.

The elliptic pairing (6.2) is the basic property for constructing a class δ_G in $K_G^m(TX)$ from the equivariant $\mathrm{spin}^c(m)$ structure on X.

For brevity set $H = \mathrm{spin}^c(m)$. We define a $G\times H$ complex of vector bundles ([5],page 489) over $P\times V$:

$$P \times V \times \Delta_+ \xrightarrow{\Phi} P \times V \times \Delta_-$$

$$\Phi(p,v,\delta) = (p,\theta(v,\delta)).$$

The $G\times H$ action on $P\times V$ is given by

$$(g,h)(p,v) = (gph^{-1},hv);$$

the action on $P \times V\times \Delta_\pm$ is given by

$$(g,h)(p,v,b) = (gph^{-1}, hv,hb).$$

Since θ is an elliptic pairing, this complex defines an element

$$\delta_G \in K^*_{G\times H}(P\times V) = K_G^m(P \times_H V) = K_G^m(TX).$$

Here are some examples in which δ_G or a close variant defines an orientation.

Example 1: X is a point, m is even, $G = U(m/2)$ is the unitary group of isometries of $C^{m/2}$ and $V = C^{m/2}$ denotes the standard $U(m/2)$ module. It is a $U(m/2)$ bundle over a point. We define an orientation class Δ_U for V.

The elliptic pairing $\theta : V \times \Delta_+ \longrightarrow V \times \Delta_-$ of (6.2) gives an elliptic $U(m/2)$ complex

$$\theta : V \times \Delta_+ \longrightarrow V \times \Delta_-$$

over a point. Here Δ_+ and Δ_- are $U(m/2)$ modules via the homomorphism ψ_0 of (6.3). This complex defines an element

$$\Delta_U \in K_U^*(V).$$

Proposition 6.13 $K_U^*(V)$ <u>is a free module over</u> $R(U) = K_U^*(p)$ <u>generated by</u> Δ_U.

Proof: The symbol of the DeRham complex of V, $\lambda_V \in K_U^*(V)$ is a generator [5], so

$$\Delta_U = a \cdot \lambda_V \in K_U^*(V)$$

for some $a \in R(U)$. Let $j : T \longrightarrow U$ denote the inclusion of the maximal torus and j' the composition

$$K_U^*(V) \xrightarrow{i^*} K_U^*(0) \xrightarrow{j^*} K_T^*(0) = R(T)$$

Here i^* is the restriction defined by the inclusion of the origin 0. Now

$$j'\lambda_V = \lambda_{-1}(V) = \prod_{i=1}^{m/2} (1-t_i) \in R(T)$$

$$R(T) = Z[t_1,t_1^{-1}, t_2,t_2^{-1}, \cdots, t_{m/2},t_{m/2}^{-1}].$$

But $j'\Delta_u = (\Delta_+ - \Delta_-)|_T = \pi(1-t_i^{-1})t_i$. This follows from the definition of ψ_0 and the fact that the trace of $t = (t_1,t_2, \cdots, t_{m/2})$ acting on Δ_+ minus the trace of t on Δ_- is $\pi(1-t_i^{-1})t_i$. Since $R(U) \longrightarrow R(T)$ is injective, $a = (-1)^{m/2}$.

If $G \xrightarrow{\rho} U(m/2)$ is a representation then $\Delta_G \in K_G^*(V)$, is defined to be $\rho^*\Delta_U$. It generates $K_G^*(V)$ freely over $R(G)$.

Example 2: X is a spin$^c(m)$ manifold with an S^1 action which satisfies (6.12).

Proposition 6.14 δ_{S^1} <u>is an orientation for X.</u>

Proof: Let 0 be an orbit of the S^1 action on X and Γ the isotropy group of any point p of 0; so $0 = S^1/\Gamma$. Let V be a complex Γ module which as a real Γ module is TX_p. If $i : 0 \longrightarrow X$ is the inclusion, the real S^1 vector bundle defined by $S^1 \times_\Gamma V$ over 0 is i^*TX and

$$i^*\delta_{S^1} \in K_{S^1}^*(i^*TX) = K_\Gamma^*(V)$$

is Δ_Γ which generates $K_\Gamma^*(V)$ freely over $K_\Gamma^*(p) = K_{S^1}^*(0)$.

There is a theorem of Stewart [11] improved by Su [12] which to our knowledge has found little use until now. It is fundamental to the rest of our discussion. The situation is this: X is a paracompact space supporting a left action of a torus group T_1; P is a principle T_2 bundle over X. The torus T_2 acts on the right of P.

Theorem 6.15 Stewart [11] and Su [12]. If $H^1(X,Z) = 0$, the left action of T_1 on X lifts to a left action of T_1 on P which commutes with the principle right action of T_2 on P. If $(t,p) \longrightarrow t\cdot p$ and $(t,p) \longrightarrow t \circ p$ denote two liftings of T_1 to P then there is a homomorphism $\theta : T_1 \longrightarrow T_2$ such that

$$t \circ p = t\cdot p\cdot\theta(t).$$

We shall restrict application of this theorem to the case $T_1 = T_2 = S^1$. Suppose, in addition to the hypothesis of (6.15), that X is a smooth manifold of dimension m such that $|X|$ has a $\mathrm{spin}^c(m)$ structure. Let P be a principle $spin^c(m)$ bundle over X associated to the tangent bundle of X, i.e., $P \times_{spin^c(m)} V = TX$.

Theorem 6.16 <u>The left S^1 action on X lifts to a left S^1 action on P which commutes with the right action of $spin^c(m)$ on P. Thus the action of X satisfies the hypothesis (6.12).</u>

Proof: Let Q denote the principle $SO(m)$ bundle associated to the tangent bundle of X. From the exact sequence of groups

$$S^1 \longrightarrow spin^c(m) \longrightarrow SO(m)$$

we see that P is a principle S^1 bundle over Q. If $m>2$, then $H^1(Q,Z) = 0$. There is a natural left S^1 action on Q commuting with the principle right action of $SO(m)$. By (6.15) this action lifts to P and commutes with the principle right action of S^1. We may assume that the lifted action in fact commutes with the principle $spin^c(m)$ action on P and covers the S^1 action on X. [9]

Corollary 6.17 <u>Let X be a smooth $spin^c(m)$ manifold with $H^1(X,Z) = 0$ and which supports a smooth S^1 action. Then the class $\delta_{S^1} \in K^*_{S^1}(TX)$ is defined and is an orientation for X.</u>

Proof: This follows from Theorem 6.16 and Proposition 6.14.

Corollary 6.18 Let X satisfy the hypothesis of Corollary 6.17. Then there are isomorphisms

$$\Delta_1^X : K^*_{S^1}(X) \longrightarrow K^*_{S^1}{}^*(TX).$$

$$\Delta^X : K^*_{S^1}(X)_P \longrightarrow K^*_{S^1}(TX)_P.$$

Theorem 6.19 Let X be a closed smooth S^1 manifold such that $|X|$ has a $spin^c(m)$ structure then the bilinear form $< >_X$ on $K^*_{S^1}(X)_P$ defined by

$$<a,b>_X = Id^X_{S^1} \otimes 1_R(\Delta^X(a)\cdot b) \in R$$

is non-degenerate, i.e., the associated R homomorphism

$$\Phi^X_R : K^*_{S^1}(X)_P \longrightarrow Hom_R(K^*_{S^1}(X)_P, R)$$

induces an isomorphism

$$\Phi'^X_R : K^*_{S^1}(X)_P/T_X \longrightarrow Hom_R(K^*_{S^1}(X)/T_X, R)$$

where T_X is the R torsion subgroup of $K^*_{S^1}(X)_P$.

Proof: Set $h^* = (K^*_{S^1})_P$. Then $\phi^X_R = \phi^X \circ \Delta^X$; so the theorem follows from (6.18) and (5.6).

REFERENCES

[1] Atiyah, M. F., K-Theory, Benjamin (1967).

[2] ____________, Vector Fields on Manifolds, Arbeitsgemeinschaft Fur Forschung Des Landes Nordrhein-Westfalen, Heft 200, (1969).

[3] Atiyah, M. F., and Segal, G., The index of elliptic operators II, Annals of Math., 87 (1968) pp. 531-545.

[4] ____________________________, Equivariant K theory and completion, J. Differential Geometry, 8 (1969) pp. 1-18.

[5] Atiyah, M. F., and Singer, I., The index of elliptic operators I and III, Annals of Math, 87 (1968) pp. 484-530 and 546-604.

[6] Bredon, G., Introduction to Compact Transformation Groups, Academic Press, (1972).

[7] Lee, C. N., Equivariant homology theories, Proceedings of the Conference on Transformation Groups, Springer-Verlag, (1968), pp. 237-244.

[8] Milnor, J., The Representation Rings of Some Classical Groups, Notes, Princeton University, (1963).

[9] Petrie, T., Smooth S^1 actions on homotopy complex projective spaces and related topics, Bull. A.M.S., Vol. 78, No. 2 (1972) pp. 105-153.

10] Segal, G., Equivariant K theory, Inst. Hautes E'tudes Sci. Publ. Math., No. 34 (1968) pp. 129-151.

11] Stewart, T. E., Lifting the action of a group in a fiber bundle, Bull. A.M.S., Vol. 66, (1960), pp. 129-132.

12] Su, J. C., Transformation groups on cohomology projective spaces, Trans. A.M.S., Vol. 106 (1963) pp. 305-318.

13] Vasquez, A., Poincare duality for $K_G(G/H)$, to appear.

14] Wasserman, A., Equivariant differential topology, Topology 8 (1969) pp. 127-150.

PART II

A SETTING FOR SMOOTH S^1 ACTIONS WITH APPLICATIONS TO REAL ALGEBRAIC ACTIONS ON $P(\mathbb{C}^{4n})$

1. Introduction and Notation.

Let G be a compact Lie Group. A smooth G manifold Y consists of a pair $(|Y|, \chi)$ where $|Y|$ is a smooth manifold and χ is a representation of G in the group Diff $(|Y|)$ of diffeomorphisms of $|Y|$ such that the map $\mu : G \times |Y| \to |Y|$ defined by $\mu(g,y) = \chi(g)[y]$ is smooth for $g \in G$, $y \in |Y|$. For brevity, write $y \in Y$ and $\chi(g)[y] = gy$. If X is a G manifold a map $f : X \to Y$ must satisfy $f(gx) = gf(x)$. The tangent space TY of a G manifold Y is a G manifold and if $p \in Y$ is a fixed point i.e. $p \in Y^G = \{y \in Y | gy = y\}$, the restriction of TY to p, written TY_p, is a real G module.

The main questions to which we address ourselves are these: Let M and N be two homotopy equivalent manifolds. (1) Suppose there is a G manifold Y with $|Y| = M$. Is there a G manifold X with $|X| = N$? (2) Given Y with $|Y| = M$ how can we construct X with $|X| = N$?
The central question which must be answered for dealing with (1) and (2) is: What are the relations among the representations of G on the tangent spaces TY_p, $p \in Y^G$ and the global invariants of $|Y|$ e.g. its Pontrjagin classes and cohomology.

For dealing with these questions, we introduce the set $S_G(Y)$ attached to the closed G manifold Y. It consists of equivalence classes of pairs (X,f) where X

is a closed G manifold and $f : X \to Y$ is a map such that

(i) $|f| : |X| \to |Y|$ is a homotopy equivalence.
Here $|f|$ means the underlying map to f obtained by neglecting its relation to G.

(ii) $|f^G| : |X^G| \to |Y^G|$ is a homotopy equivalence.
Two pairs (X_i, f_i) $i = 0,1$ are equivalent if there is a G homotopy equivalence $\phi : X_0 \to X_1$ such that $f_1\phi$ is G homotopic to f_0. The equivalence class of (X,f) is denoted by $[X,f] \in S_G(Y)$. The identity map of Y is denoted by I_Y and the element $[Y,I_Y] \in S_G(Y)$ is called the trivial element.

Condition (i) imposes stringent restrictions on $\ker f^*$ and $\operatorname{coker} f^*$ where f^* is the induced map in equivariant K theory. The essential algebraic fact here is the Localization Completion Lemma 3.3. When $G = S^1$, we connect the algebras $K^*_{S^1}(X)$ and $K^*_{S^1}(Y)$ with the S^1 representations $\{TX_p \mid p \in X^{S^1}\}$ and $\{TY_q \mid q \in Y^{S^1}\}$ and with the Pontrjagin classes of $|X|$ and $|Y|$. The invariant of $[X,f] \in S_{S^1}(Y)$ which gives these connections is the torsion (see 2.4)

$$f_*(1_X) \in \tilde{K}^*_{S^1}(Y)$$

If the torsion is the identity of the algebra $\tilde{K}^*_{S^1}(Y)$, then $TX_p = TY_{f(p)}$ $\forall\, p \in X^{S^1}$ and $|f|^*$ preserves Pontryjagin classes provided $|Y|$ satisfies suitable hypotheses. In particular if f is an S^1 homotopy equivalence, $|f|^*$ preserves Pontryjagin classes (Theorem 6.4).

As an application of the preceding ideas, we study S^1 actions on $P(\mathbb{C}^n)$, the space of complex lines in $\mathbb{C}^n$. In particular for certain S^1 modules Ω we produce a non trivial element $[X(\omega),\bar{\omega}] \in S_{S^1}(P(\Omega))$ and illustrate the necessity of the hypothesis in the theorems comparing the local and global properties of X and Y when $[X,f] \in S_{S^1}(Y)$. The S^1 manifolds $X(\omega)$ also show the necessity of the hypothesis of Bredon [8] concerning the structure of fixed points sets of subgroups of S^1.

2. Specifics about S^1.

The complex representation ring of S^1 $R(S^1)$ is $Z[t,t^{-1}]$. If $\lambda_1, \ldots \lambda_n$ are integers, then $t^{\lambda_1} + \ldots + t^{\lambda_n} \in Z[t,t^{-1}]$ represents a complex S^1 module Γ of dimension n and $\lambda_{-1}(\Gamma) = \Sigma(-1)^i \lambda^i(\Gamma) \in R(S^1)$ is the product $\prod_{i=1}^{n} (1 - t^{\lambda_i})$. Here $\lambda^i(\Gamma)$ is the ith exterior power of Γ.

The ring $Z[t,t^{-1}]$ contains two important sets of prime ideals P and $P_1 \subset P$.

2.1 P is the set of prime ideals defined by cyclotomic polynomials $\phi_m(t)$ as m ranges over all integers.

2.2 P_1 is the set of prime ideals defined by cyclotomic polynomials $\phi_{p^r}(t)$ where p is a prime integer and r is arbitrary; i.e. each integral prime p and positive integer r contribute a prime ideal to P_1.

If $\mathcal{S}$ is any set of prime ideals in $R(S^1)$ and A is an $R(S^1)$ module, $A_{\mathcal{S}}$ denotes A localized at $\mathcal{S}$. The basic ring over which all our algebra takes place is the ring $R = R(S^1)_P$. It is a principle ideal domain whose field of fractions we denote by F. The only prime ideals of R are those in P. Each $\rho \in P$ defines a valuation on R and a norm $\|\ \|_\rho$ on F written $x \to \| x \|_\rho$, $x \in F$. In order to be explicit, if $^a/_b \in F$ and $\rho = (\phi_m(t))$, then $\| ^a/_b \|_\rho = \phi_m(t)^k$ where $\phi_m(t)^\alpha \| a$, $\phi_m(t)^\beta \| b$ and $k = \alpha - \beta$.

Let M and N be two complex S^1 modules of the

same dimension and $\rho \in P$. We say that M and N are equivalent at ρ if $\lambda_{-1}(M) \neq 0$, $\lambda_{-1}(N) \neq 0$ and $\left\| \lambda_{-1}(M) / \lambda_{-1}(N) \right\|_\rho = 1$. We write this as $M \underset{\rho}{=} N$. If M and N are real S^1 modules of the same even dimension, choose complex S^1 modules $\bar{M}$ and $\bar{N}$ whose underlying real S^1 modules are M and N and define $M \underset{\rho}{=} N$ if $\bar{M} \underset{\rho}{=} \bar{N}$. Because $\lambda_{-1}(\bar{M})$ and $\lambda_{-1}(\bar{N})$ are products of cyclotomic polynomials, it follows that if $M \underset{\rho}{=} N$ for all $\rho \in P$, then $M = N$.

Hypothesis H: (i) Y is a closed d dimensional S^1 manifold and Y^{S^1} consists of isolated fixed points, (ii) $|Y|$ is a spin^c manifold (§4) (iii) the odd dimensional rational cohomology of $|Y|$ is zero, (iv) $H^1(|Y|, Z) = 0$.

When Y satisfies H(i) and $q \in Y^{S^1}$, TY_q is a real d dimensional S^1 module and there are integers $y_i(q)$ $i = 1,2,\ldots d/2$ such that $\overline{TY}_q = \Sigma_{i=1}^{d/2} t^{y_i(q)} \in R(S^1)$.

Denote by $K^*_{S^1}(Y)$ the complex equivariant K theory of Y. It is a module over $R(S^1)$. If $\gamma \in K^0_{S^1}(Y)$ and $p \in Y^{S^1}$, γ_p or $\gamma(p)$ will denote the restriction of γ to p. It is an element of $K^0_{S^1}(p) = R(S^1)$; so is a function of t. The value at t is denoted either by $\gamma_p(t)$ or $\gamma(p)[t]$. Denote by $\varepsilon_Y\colon K^*_{S^1}(Y) \to K^*(|Y|)$ the $R(S^1)$ homomorphism obtained by taking an S^1 vector bundle η over Y to its underlying vector bundle $|\eta|$ over $|Y|$. When Y is a point, this is the augmentation homomorphism $\epsilon : Z[t,t^{-1}] \to Z$. The latter induces an augmentation $\tilde{\epsilon} : R \to Q$.

When Y is a closed S^1 manifold, we have the Atiyah-Singer Index homomorphism

$$\text{Id}^Y_{S^1} : K^*_{S^1}(TY) \to R(S^1) \ [2].$$

f Y satisfies $H(ii)\&(iv)$, there is an isomorphism

$$\Delta^Y : K^*_{S^1}(Y) \to K^*_{S^1}(TY) \ [11].$$

et us denote by $\tilde{K}^*_{S^1}(Y)$ the R module $K^*_{S^1}(Y)_P/T_Y$

where T_Y is the R torsion submodule of $K^*_{S^1}(Y)_P$.

The homomorphisms $\text{Id}^Y_{S^1}$ and Δ^Y induce R homomorphisms

loosely called $\text{Id}^Y_{S^1}$ and Δ^Y)

$$\text{Id}^Y_{S^1} : \tilde{K}^*_{S^1}(TY) \to R \quad \text{and} \quad \Delta^Y : \tilde{K}^*_{S^1}(Y) \to \tilde{K}^*_{S^1}(TY).$$

Define an R valued bilinear form $< >_Y$ on $\tilde{K}^*_{S^1}(Y)$ by

$$<a,b>_Y = \text{Id}^Y_{S^1}(\Delta^Y(a).b) \quad \text{for} \quad a,b \ \epsilon \ \tilde{K}^*_{S^1}(Y) .$$

Theorem 2.3 (Part I, 6.19). If Y satisfies $H(ii)\&(iv)$, the bilinear form $< >_Y$ is non singular i.e. defines an isomorphism of $\tilde{K}^*_{S^1}(Y)$ with $\text{Hom}_R(\tilde{K}^*_{S^1}(Y),R)$.

Observe that if Y satisfies $H(ii)\&(iv)$ and $X,f] \ \epsilon \ S_{S^1}(Y)$, then X does also. In fact we choose the pin^c structure on $|X|$ induced by the map $|f|$. In view f this discussion, the bilinear form $< >_X$ is defined hen $< >_Y$ is defined. This means that for elements

$[X,f] \in S_{S^1}(Y)$, we can define an induction homomorphism

$$f_* : \tilde{K}_{S^1}(X) \to \tilde{K}^*_{S^1}(Y) \quad \text{by}$$

2.4 $$\langle f_*(x), y\rangle_Y = \langle x, f^*(y)\rangle_X \ .$$

It is an R homomorphism which satisfies

2.5 $$f^* f_*(x) = f^* f_*(1_X) . x$$

where $1_X \in \tilde{K}^*_{S^1}(X)$ is the identity [16]. The element

$$f_*(1_X) \in \tilde{K}^*_{S^1}(Y)$$

is called the torsion of $[X,f]$.

Remark: To connect the local and global geometry we are forced to deal with $\tilde{K}^*_{S^1}(Y)$ rather than $K^*_{S^1}(Y)$; hence, we ignore the fact that the induction homomorphism can be geometrically defined from $K^*_{S^1}(X)$ to $K^*_{S^1}(Y)$.

3. The relation between completion and localization.

Let G be a compact Lie Group and $\varepsilon_G : R(G) \to Z$ the augmentation homomorphism from the complex representation ring of G (character ring) to the integers. The kernel of this homomorphism is the augmentation ideal $I \subset R(G)$. The completion of $R(G)$ with respect to I is the ring $\widehat{R(G)}$ which is

$$\widehat{R(G)} = \lim_{\leftarrow} R(G)/I^n .$$

We want to relate the operation of completion $M \to \hat{M}$ to localization $M \to M_\rho$ at certain prime ideals $\rho \in R(G)$. Here M is an $R(G)$ module. The prime ideals we have in mind are those defined by the cyclic subgroups of G of prime power order. Let p be a prime and ξ denote a primitive p^r th root of 1. Let $r : Z[\xi] \to Z/pZ$ denote the ring homomorphism defined by $r(\xi) = 1 \bmod p$. Observe that if $g \in G$ is an element of order p^r and $\chi \in R(G)$, $\chi(g) \in Z[\xi]$; also, if $e \in G$ denotes the identity, $\chi(e) \in Z \subset Z[\xi]$.

Lemma 3.1 Let $g \in G$ be an element of order p^r. Then $r\chi(g) = r\chi(e)$.

Proof: We may assume that G is the cyclic group generated by g denoted by Z_{p^r}. Then

$$R(G) = Z[t]/(1-t^{p^r})$$

and under this identification we view $\chi \in R(G)$ as an element

$\chi(t) \varepsilon\ ^{Z[t]}/_{(1-t^{p^r})}$ whose value at g^k is obtained by evaluating $\chi(t)$ at ξ^k.

Thus if $\chi(t) = \Sigma a_i t^i$, $\chi(g) - \chi(1) = \Sigma a_i(\xi^i - 1) \equiv 0(\xi-1)$ and $r\chi(g) = r\chi(1)$.

Localization-Completion Lemma 3.2 <u>Let</u> $g \varepsilon G$ <u>be an element of prime power order. Let</u> $\rho = \{\chi \varepsilon R(G) | \chi(g) = '0\}$. <u>If</u> M <u>and</u> N <u>are two</u> $R(G)$ <u>modules and</u> $f : M \to N$ <u>is an</u> $R(G)$ <u>homomorphism such that</u> $\hat{f} : \hat{M} \to \hat{N}$ (<u>the completion of</u> f) <u>is an isomorphism, then</u> f_ρ (<u>the lozalization of</u> f) <u>is an isomorphism</u>.

Proof: Since localization and completion are exact functors, it suffices to show that if M is an $R(G)$ module such that $\hat{M} = 0$, then $M_\rho = 0$. Again since localization and completion are exact functors, we can reduce to the case where M is cyclic over $R(G)$, i.e., $M = {}^{R(G)}/_J$ where $J = (\chi_1, \ldots \chi_n)$ is an ideal in $R(G)$ with generators $\chi_i \varepsilon \rho$. If $\hat{M} = 0$, then $\hat{J} = \widehat{R(G)}$ and there are elements a_i $i = 1,2 \ldots n$ in $\widehat{R(G)}$ such that in $\widehat{R(G)}$, $\sum_{i=1}^{n} a_i\chi_i = 1$. Apply the augmentation homomorphism in $\widehat{R(G)}$ to this equation. Then in Z we have $\sum_{i=1}^{n} a_i(e)\chi_i(e) = 1$. But by Lemma 3.1 $r\chi_i(e) = r\chi_i(g) = 0$ so $\chi_i(e) \equiv 0 \bmod p$ which contradicts the above equation.

The Localization Completion Lemma has interesting applications for comparing the "mod p^r" homological aspects of S^1 manifolds X and Y when $[X,f] \varepsilon S_{S^1}(Y)$. For xample, we can compare $X^{Z_{p^r}}$ and $Y^{Z_{p^r}}$, $K^*_{S^1}(X)_\rho$ and $K^*_{S^1}(Y)_\rho$ and the representations of S^1 on the normal undles to $X^{S^1} \subset X$ and $Y^{S^1} \subset Y$ at primes $\rho \in P_1$.

heorem 3.3 Let X and Y be G manifolds and $f : X \to Y$ G map such that $|f| : |X| \to |Y|$ is a homotopy equivalence. et $g \in G$ be an element of prime power order and ρ the rime ideal of characters of $R(G)$ which vanish at g. Then $f^*_\rho : \tilde{K}^*_G(Y)_\rho \to \tilde{K}^*_G(X)_\rho$ is an isomorphism.

roof: It follows from the Atiyah-Segal Completion Theorem [6] hat $\hat{f}^* : \hat{K}^*_G(Y) \to \hat{K}^*_G(X)$ is an isomorphism. y the Localization Completion Lemma 3.3 $f^*_\rho : K^*_G(Y)_\rho \to K^*_G(X)_\rho$ s an isomorphism. This readily implies $f^*_\rho : \tilde{K}^*_G(Y)_\rho \to \tilde{K}^*_G(X)_\rho$ is an isomorphism.

orollary 3.4 Assume the hypothesis of Theorem 3.3. f G is S^1 and $Z_{p^r} \subset G$ is the subgroup generated by ε G, $(f^{Z_{p^r}})^*_\rho : \tilde{K}^*_G(Y^{Z_{p^r}})_\rho \to \tilde{K}^*_G(X^{Z_{p^r}})_\rho$ is an isomorphism.

roof: This is an easy consequence of Theorem 3.3 and the ocalization theorem of [5] .

4. Remarks on $spin^c$ structures.

Let d be an even integer. The group $spin^c(d)$ has center $SO(2)$ and its quotient by the center is $SO(d)$ (Part I §6). Let M be an oriented manifold and M_1 the total space of the principle right $SO(d)$ bundle associated to TM. If there is a principle right $spin^c(d)$ bundle over M with total space M_2 such that $M_{2/SO(2)} = M_1$ as a principle $SO(d)$ bundle over M, then we say that M is a $spin^c$ manifold with $spin^c$ structure determined by M_2.

Suppose that Y is a smooth S^1 manifold with underlying space $|Y| = M$. There is a canonical S^1 action on M_1 commuting with the principle $SO(d)$ action on M_1 and covering the S^1 action on Y [11] p.117, 4.7. We denote the resulting S^1 manifold by Q. We say that the S^1 action on Y preserves the $spin^c$ structure if there is a smooth S^1 manifold P with $|P| = M_2$ such that the natural projection $M_2 \to M_1$ is S^1 equivariant i.e. defines a map $P \to Q$. In this situation we obtain an S^1 line bundle ω^Y over Y with total space $P \times_{spin^c(d)} \mathbb{C}$. Here $spin^c(d)$ acts on $\mathbb{C}$ via the canonical homomorphism of $spin^c(d)$ to U_1 with kernel $spin(d)$. If $q \in Y^{S^1}$, ω^Y_q is a complex one dimensional S^1 module whose character is $\omega^Y_q(t) = t^{\omega^q}$ for some integer ω^q.

The half spin representations give rise to a class $\delta^Y_{S^1} \in K^0_{S^1}(TY)$ which generates $K^*_{S^1}(TY)$ freely as a $K^*_{S^1}(Y)$ module. In fact the isomorphism Δ^Y is defined by $\Delta^Y(y) = \delta^Y_{S^1} \cdot y$ for $y \in K^*_{S^1}(Y)$ [11]. Moreover if

$\delta^Y_{S^1}(q) \in R(S^1)$ denotes the restriction of $\delta^Y_{S^1}$ to an isolated fixed point q of Y^{S^1}, then

$$4.1 \qquad \frac{\lambda_{-1}(TY_q \otimes \mathbb{C})(t)}{\delta^Y_{S^1}(q)(t)} = t^{-\omega^q/2} \prod_{i=1}^{d/2} \left(t^{\frac{-y_i(q)}{2}} - t^{\frac{y_i(q)}{2}}\right)$$

where $\overline{TY}_q = \Sigma_{i=1}^{d/2} t^{y_i(q)/2}$. See [11] p.124.

If Y satisfies H(ii) & (iv), we may assume that the S^1 action on Y preserves the spin^c structure and likewise for X if $[X,f] \in S_{S^1}(Y)$ (Part I§6). Moreover, we assume that f preserves spin^c structures which means that $|f|^*|\omega^Y| = |\omega^X|$.

Proposition 4.2 <u>There is an integer</u> b <u>such that</u> $f^*\omega^Y = t^b\omega^X$ <u>in</u> $K^0_{S^1}(X)$.

Proof: The line bundles $|f^*\omega^Y| = |f|^*|\omega^Y|$ and $|\omega^X|$ are equal. The result is now an easy consequence of [11] p.126, Theorem 6.1.

Corollary 4.3 <u>There is an integer</u> b <u>such that for every</u> $p \in X^{S^1}$, $q = f(p)$, $\omega^q = \omega^p + b$ <u>where</u> $\omega^Y_q(t) = t^{\omega^q}$, $\omega^X_p(t) = t^{\omega^p}$.

The class in $H^2(|Y|,Z)$ determined by the spin^c structure is denoted by $c = c_{|Y|}$. It is the first Chern class of $|\omega^Y|$. The fundamental classes of $|X|$ and $|Y|$ are denoted by $[X]$ and $[Y]$ in their respective homologies.

Since the spin^c structure on $|X|$ is compatible with $|f|$, we have $c_{|X|} = |f|^* c_{|Y|}$. Let $A(|X|)$ denote the cohomology class in $H^*(|X|,Q)$ associated to $|TX|$ by the power series $x_{/2}(\sinh x_{/2})^{-1}$ [4]; so $A(|X|)$ is a polynomial in the Pontrjagin classes of $|X|$. Let $g : |Y| \to |X|$ be a map which is a homotopy inverse of $|f|$. Since the term of degree zero of the classes $A(|X|)$ and $A(|Y|)$ is 1, they are units in their respective cohomologies and

$$4.4 \qquad g^*A(|X|)/_{A(|Y|)} = |X|/_{|Y|}$$

is a unit of $H^*(|Y|,Q)$ which we call the relative differential structure of $|X|$ and $|Y|$. If $|X|/_{|Y|}$ is 1, then $|f|^*$ preserves Pontrjagin classes. Note that $A(|Y|)$ determines the Pontrjagin classes of $|Y|$.

In addition to the local relation between the class $\delta^Y_{S^1}$ and the representations of S^1 on the tangent spaces of Y at isolated fixed points given by 4.1, we need the following global relation which brings in the Pontrjagin classes of $|Y|$. If $\mathrm{Id}^{|Y|} : K^0(|TY|) \to Z$ denotes the Atiyah-Singer homomorphism for the trivial group, we have

$$4.5 \qquad \mathrm{Id}^{|Y|}(\varepsilon_{TY}(\delta^Y_{S^1}).u) = \langle A(|Y|)e^{c/2} . ch(u), [Y]\rangle$$

for $u \in K^*(|Y|)$ and for any $v \in K^0_{S^1}(TY)$, $1 \in S^1$,

$$4.6 \qquad \varepsilon \mathrm{Id}^Y_{S^1}(v) = \mathrm{Id}^Y_{S^1}(v)[1] = \mathrm{Id}^{|Y|}(\varepsilon_{TY}(v)) .$$

ee [11] p.125 and [2] p.501. The expression on he righthand side of 4.5 denotes evaluation of the ndicated cohomology class of $|Y|$ on the fundamental lass $[Y] \in H_d(|Y|)$.

5. The induction homomorphism.

The normal bundle NY^{S^1} of Y^{S^1} can be given the structure of a complex S^1 vector bundle. This implies that $|NY^{S^1}|$ has a spin^c structure and if $|Y|$ has a spin^c structure, so does $|Y^{S^1}|$. In particular the bilinear form $< \ >_{Y^{S^1}}$ is defined and non singular (The condition $H^1(|Y^{S^1}|,Z) = 0$ is not needed.) and we can associate to the inclusion $i_Y : Y^{S^1} \to Y$ the induction homomorphism $(i_Y)_*$ and we set $(i_Y)_*(1_{Y^{S^1}}) = \alpha_Y \in \tilde{K}^*_{S^1}(Y)$. If $[X,f] \in S_{S^1}(Y)$, $|f^{S^1}|$ is a homotopy equivalence so $(f^{S^1})^* : \tilde{K}_{S^1}(Y^{S^1}) \to \tilde{K}_{S^1}(X^{S^1})$ is an isomorphism which we use to identify these two algebras.

Proposition 5.1 f^*, i_X^* and i_Y^* <u>are monomorphisms which induce isomorphisms over</u> F.

Proof: It is an immediate consequence of the Localization Theorem [5] that these homomorphisms induce isomorphisms over F. Since all algebras are free of R torsion they are monomorphisms.

The connection between $f_*(1_X)$ and the representations TX_p and $TY_{f(p)}$ is provided by

Proposition 5.2 <u>If</u> Y <u>satisfies</u> H(ii)&(iv)

(a) $i_Y^* f_*(1_X) = i_Y^*(\alpha_Y) / i_X^*(\alpha_X)$. <u>If in addition</u> H(i) <u>holds</u>, $p \in X^{S^1}$ <u>and</u> $q = f(p)$;

b) $i_Y^*(\alpha_Y)_q(t) = \lambda_{-1}(TY_q \otimes \mathbb{C})(t) / \delta^Y_{S^1}(q)(t)$

c) $$i_Y^* f_*(1_X)_q(t) = t^{\frac{\omega^p - \omega^q}{2}} \prod_{i=1}^{d/2} \frac{(t^{-y_i(q)/2} - t^{y_i(q)/2})}{(t^{-x_i(p)/2} - t^{x_i(p)/2})} =$$

$\mu_q \, \lambda_{-1}(\overline{TY}_q)/\lambda_{-1}(\overline{TX}_p)$ <u>and</u> $\mu_q \in R$ <u>is a unit</u>.

roof: (a) and (b) are results of [16]. (c) follows rom (a),(b) and (4.1).

heorem 5.3 <u>Suppose</u> Y <u>satisfies</u> H(ii)&(iv). <u>Let</u> $\mathcal{S}$ <u>e a set of prime ideals of</u> R. <u>If</u> $f^*_{\mathcal{S}}$ <u>is an isomorphism,</u> $_*(1_X)$ <u>is a unit of</u> $\tilde{K}^*_{S^1}(Y)_{\mathcal{S}}$. <u>This hypothesis is</u> <u>atisfied for</u> $\mathcal{S} = P_1$. <u>If</u> H(i) <u>also holds and if</u> $_*(1_X)$ <u>is a unit of</u> $\tilde{K}^*_{S^1}(Y)_{\mathcal{S}}$, <u>then</u> $TX_p \underset{\rho}{=} TY_{f(p)}$ <u>for</u> <u>ll</u> $\rho \in \mathcal{S}$, $p \in X^{S^1}$.

roof: If $f^*_{\mathcal{S}}$ is an isomorphism, then $(f_*)_{\mathcal{S}}$ is an somorphism because the bilinear forms $<\ \ >_X$ and $<\ \ >_Y$ re non degenerate. If $x \in \tilde{K}_{S^1}(X)_{\mathcal{S}}$, then

$^*_{\mathcal{S}}\ (f_*)_{\mathcal{S}}\ (x) = [f^*_{\mathcal{S}}\ (f_*)_{\mathcal{S}}\ (1_X)] \cdot x$ (2.5).

ince $f^*_{\mathcal{S}}(f_*)_{\mathcal{S}}$ is an isomorphism over $R_{\mathcal{S}}$, $f^*_{\mathcal{S}}(f_*)_{\mathcal{S}}(1_X)$ s a unit of $\tilde{K}^*_{S^1}(X)_{\mathcal{S}}$. Since $f^*_{\mathcal{S}}$ is an algebra somorphism, $(f_*)_{\mathcal{S}}(1_X) = f_*(1_X)$ is a unit of $\tilde{K}^*_{S^1}(Y)_{\mathcal{S}}$. he second statement follows from (3.3).

If $f_*(1_X)$ is a unit of $\tilde{K}^*_{S^1}(Y)_{\mathcal{S}}$, $i_Y^* f_*(1_X)_q$ is a nit of $R_{\mathcal{S}}$. Then $\| i_Y f_*(1_X)_q \|_\rho = 1$ for $\rho \in \mathcal{S}$. From 5.2 c), $TX_p \underset{\rho}{=} TY_{f(p)}$.

Remark: We can dispense with $H(i)$ by replacing TX_p and $TY_{f(p)}$ by $NX_p^{S^1}$ and $NY_{f(p)}^{S^1}$.

Suppose that N and M are complex S^1 vector bundles over Y such that $N^{S^1} = M^{S^1} = Y^{S^1}$. If $F' : N \to M$ is a proper fiber preserving map, then $\lambda_{-1}(M)/\lambda_{-1}(N) \in \tilde{K}^*_{S^1}(Y)$. Suppose that F' is properly S^1 homotopic to F where F is transverse regular to Y. Let $X = F^{-1}(Y)$ and $f : X \to Y$ be the composition $X \subset N \to Y$.

Proposition 5.4 $f_*(1_X) = \lambda_{-1}(M)/\lambda_{-1}(N)$

Proof: The hypothesis imply that $X^{S^1} = Y^{S^1}$ and $TX \oplus f^*(M) = f^*(TY \oplus N)$. Hence,

$$NX^{S^1} \oplus (f^{S^1})^*(i_Y^* M) = (f^{S^1})^*(NY^{S^1} \oplus i_Y^* N) .$$

Now $\alpha_X = (i_X)_*(1_{X^{S^1}})$ and $(i_X)^*(i_X)_*(u) = \lambda_{-1}(NX^{S^1}).u$, [2] p.497; so using $(f^{S^1})^*$ as an identification, we have by (5.2 a) $i_Y^* f_*(1_X) = i_Y^*(\alpha_Y)/i_X^*(\alpha_X) = i_Y^*(\lambda_{-1}(M)/\lambda_{-1}(N))$.

But i_Y^* is a monomorphism (5.1).

6. Differential Structure and the representations TX_P

Throughout §6 we assume that Y satisfies $H(i)$-(iv) and $[X,f] \in S_{S^1}(Y)$.

If $K^*_{S^1}(Z)_P$ has no $(t-1)$ torsion, the homomorphism ε_Z will induce a homomorphism $\tilde{\varepsilon}_Z : \tilde{K}^*_{S^1}(Z) \to K^*(Z) \otimes Q$.

Proposition 6.1 $K^*_{S^1}(Y)_P$ has no $(t-1)$ torsion and $\tilde{\varepsilon}_Y$ is defined and surjective.

Proof: The hypothesis H and the completion theorem of [6] together with a spectral sequence argument imply that the completion of the cokernel of $\varepsilon_Y : K^0_{S^1}(Y) \to K^0(Y)$ is finite; but coker ε_Y is complete. The proof is finished by examining the exact sequence [11] p.133.

$$\begin{array}{ccc} K^*_{S^1}(Y)_P & \xrightarrow{t-1} & K^*_{S^1}(Y)_P \\ & \nwarrow \qquad \swarrow \varepsilon_Y \otimes Q & \\ & K^*(Y) \otimes Q & \end{array}$$

In order to connect $f_*(1_X)$ and $|X|/_{|Y|}$, we introduce the homomorphism $\phi_Y = ch \circ \tilde{\varepsilon}_Y$ where ch is the Chern character homomorphism.

Theorem 6.2 $\phi_Y f_*(1_X) = |X|/_{|Y|}$

Proof: Recall that $\tilde{\varepsilon}$ is the homomorphism from R to Q induced by ε. If $y \in \tilde{K}^*_{S^1}(Y)$, then from 4.5 and 4.6 we have

$$\tilde{\varepsilon}<f_*(1_X),y>_Y \;=\; Id^Y_{S^1}(\Delta^Y(f_*(1_X).y)\;(1) \;=$$

$$Id^{|Y|}(\tilde{\varepsilon}_{TY}(\delta^Y_{S^1}) \;.\; \tilde{\varepsilon}_Y(f_*(1_X).y)) \;=$$

$$<A(|Y|)e^{c/2}\;\phi_Y(f_*(1_X).y),\;\;[Y]>$$

On the other hand the same discussion for X shows that

$$\varepsilon\;<1_X,f^*(y)>_X \;=$$
$$<|f|^*g^*A(|X|)e^{|f|^*c/2}\;\phi_X f^*(y),\;\;[X]> \;=$$
$$<g^*A(|X|)e^{c/2}\phi_Y(y),\;\;[Y]>\;.$$

From (2.4) we conclude that

$$<\phi_Y f_*(1_X)\;.\;A(|Y|)e^{c/2}\phi_Y(y),\;\;[Y]> \;=$$

$<g^*A(|X|)e^{c/2}\phi_Y(y),\;\;[Y]>$ and since ϕ_Y is surjective, $\phi_Y f_*(1_X).A(|Y|) \;=\; g^*A(|X|)$.

Lemma 6.3 <u>If</u> $f_*(1_X)$ <u>is a unit of</u> $\tilde{K}^*_{S^1}(Y)$, <u>then there is an even integer</u> b <u>such that for each</u> $q\;\varepsilon\;Y^{S^1}$, $i^*_Y f_*(1_X)_q(t) = \pm t^{b/2}$.

Proof: If $f_*(1_X)$ is a unit, then $i^*_Y f_*(1_X)_q$ is a unit of $R = \tilde{K}^0_{S^1}(p)$; so $\|i^*_Y f_*(1_X)_q\|_\rho \;=\; 1$ for all $\rho\;\varepsilon\;P$ and from the form of $i^*_Y f_*(1_X)_q(t)$ (4.3) and (5.2 c), we see that this must be $\pm t^{b/2}$. Clearly b must be even because $t^{b/2}\;\varepsilon\;R$.

Theorem 6.4 <u>If</u> f^* <u>is an isomorphism, then</u> $TX_p = TY_{f(p)}$ <u>for all</u> $p\;\varepsilon\;X^{S^1}$ <u>and</u> $|X|/_{|Y|} = 1$. <u>In particular if</u> f

s an S^1 *homotopy equivalence,* $|f|^*$ *preserves Pontrjagin classes.*

Proof: The first statement is a consequence of (5.3) which implies also that $f_*(1_X)$ is a unit. By (6.3) there is an even integer b such that $i_Y^*(t^{-b/2} f_*(1_X))_q(t) = \pm 1$ for all $q \in Y^{S^1}$. Since i_Y^* is a monomorphism by (5.1), $t^{-b/2} f_*(1_X)$ is a multiplicative element of order 2; hence, $\phi_Y(t^{-b/2} f_*(1_X))$ is a multiplicative element of order 2 in $H^*(|Y|, Q)$ so must be 1 or -1. Since $\phi_Y(f_*(1_X)) = \phi_Y(t^{b/2} f_*(1_X))$ is a class whose degree zero term is 1, we must have $\phi_Y(f_*(1_X)) = 1 = |X|/_{|Y|}$.

Remark: Hypothesis H(i) is not needed for (6.1) and (6.2)

7. The topology of some real algebraic varieties.

Let $\mathbb{H}$ denote the quaternions, $\bar{z}$ the conjugate of $z \in \mathbb{H}$ and $\mathbb{H}^n$ the n fold carteasan product of $\mathbb{H}$. The Hermitian inner product on $\mathbb{H}^n$, $\langle u,v\rangle = \Sigma\, \bar{u}_i v_i$, $u = (u_1, \ldots u_n)$, $v = (v_1, \ldots v_n)$ is preserved by Sp_n i.e. $\langle gu,gv\rangle = \langle u,v\rangle$ for $g \in Sp_n$; moreover, Sp_n acts on $\mathbb{H}^n \times \mathbb{H}^n \times \mathbb{H}$ via $g(u,v,z) = (gu,gv,z)$ and $S(\mathbb{H}^n \times \mathbb{H}^n) \times \mathbb{H}$ and $D(\mathbb{H}^n \times \mathbb{H}^n) \times \mathbb{H}$ are invariant submanifolds. Here $S(\mathbb{H}^n \times \mathbb{H}^n)$ and $D(\mathbb{H}^n \times \mathbb{H}^n)$ respectively denote unit sphere and unit disk of $\mathbb{H}^n \times \mathbb{H}^n$.

Let $w : \mathbb{H} \to \mathbb{H}$ be a smooth function. The subset

$$Z_w = \{(u,v,z) \in S(\mathbb{H}^n \times \mathbb{H}^n) \times \mathbb{H} \mid 2\langle u,v\rangle = w(z)\}$$

is invariant under the action of Sp_n. The orbit space Z_w/Sp_n is $D_w = \{(z,t) \in \mathbb{H} \times \mathbb{R} \mid |w(z)|^2 + t^2 \leq 1\}$ if $n > 1$ and is ∂D_w if $n = 1$. In fact the orbit map $\rho : Z_w \to D_w$ (∂D_w) is defined by $\rho(u,v,z) = (z, |u|^2 - |v|^2)$.

We suppose that w has these properties: $w^{-1}(0) = 0$, the degree of w is 1 and the function $z \to |w(z)|^2$ has no critical values in the half open interval $(0,1]$. Clearly the identity map I has these properties and Z_I is Sp_n diffeomorphic to $S(\mathbb{H}^n \times \mathbb{H}^n)$ and D_I is the five disk.

Lemma 7.1 <u>D_w is homeomorphic to</u> $K(\partial D_w)$ <u>the cone on</u> ∂D_w.

Proof: Let $f : \mathbb{H} \times \mathbb{R} \to \mathbb{R}$ be a smooth function which agrees with $|w(z)|^2 + t^2$ in D_w and is zero outside some compact set in $\mathbb{H} \times \mathbb{R}$. Let ξ denote the vector field $-\operatorname{grad} f$. Since ξ vanishes outside a compact set, there is a one parameter group of diffeomorphisms ϕ_s generated by ξ

9] p.10. Define the homeomorphism $h : K(\partial D_w) \to D_w$
y $h(\theta,s) = \phi_{\frac{s}{1-s}}(\theta)$ $\theta \in \partial D_w$, $s \in [0,1]$.

emma 7.2 <u>The critical points of the function</u>
$(z,t) = t$ <u>on</u> ∂D_w <u>are</u> $(0,\pm 1)$.

roof: Since $t = \pm\sqrt{1-|w(z)|^2}$ on ∂D_w, (z,t) is a
ritical point of λ iff z is a critical point of $|w(z)|^2$.
ince $0 \leq |w(z)|^2 \leq 1$, the properties of w give the
esult.

orollary 7.3 ∂D_w <u>is homeomorphic to</u> S^4 <u>and</u> D_w
<u>o</u> D^5 .

roof: ∂D_w is a smooth four manifold (7.4) supporting a
unction with exactly 2 critical points (not necessarily
on-degenerate). By Milnor [9] p.25, ∂D_w is homeomorphic
o S^4. Since D_w is the cone on ∂D_w, the proof is
omplete.

Let $\Theta : Z_w \to Z_I$ be the Sp_n map defined by
$(u,v,z) = (u,v,w(z))$. The induced map on orbit spaces
s $\bar{\Theta}$ ($\bar{\Theta}$ on D_w and $\partial\bar{\Theta}$ on ∂D_w). In fact these maps
re the restrictions of $\psi : \mathbb{H} \times R \to \mathbb{H} \times R$ defined by
$(z,t) = (w(z),t)$.

emma 7.4 ψ <u>is transverse regular to</u> ∂D_I; <u>so</u>
$D_w = \psi^{-1}(\partial V_I)$ <u>is a smooth manifold and</u> $\partial\bar{\Theta}$ <u>is a degree</u>
<u>ne map</u>.

roof: This is an easy check using the properties of w.

Theorem 7.5 <u>There is an</u> Sp_n <u>homeomorphism</u>

$S : Z_W \to Z_I = S(\mathbb{H}^n \times \mathbb{H}^n)$.

Proof: Both Z_W and Z_I are Sp_n manifolds with 2 types of isotropy groups Sp_{n-1} and Sp_{n-2}. Both orbit spaces D_W and D_I are homeomorphic to D^5 with singular orbits corresponding to points of ∂D_W resp. ∂D_I. The Sp_n map $\vdash\!\Theta\!\dashv$ induces a homotopy equivalence $\partial\bar{\vdash\!\Theta\!\dashv} : \partial D_W \to \partial D_I$ by (7.3) and (7.4). An easy calculation shows that $\bar{\vdash\!\Theta\!\dashv}$ is a homotopy equivalence. As a special case $\bar{\vdash\!\Theta\!\dashv}' : Z_W^{Sp_{n-1}} \to Z_I^{Sp_{n-1}}$ is a homotopy equivalence. (Here $Sp_1 \times Sp_{n-1} \subset Sp_n$). Since $\vdash\!\Theta\!\dashv'$ is an Sp_1 map which is a homotopy equivalence, the classification of [7] p.252 provides the existence of S.

Remark: If we had proved that D_W was diffeomorphic to D^5, we could replace homeomorphism by diffeomorphism in Theorem 7.5.

8. Real algebraic actions on P(^{4n}).

Let $T_i = S^1$ for $i = 1,2$ and let T^2 be the two torus $T_1 \times T_2$. Coordinates of a point of T^2 are (η, t). Complex Sp_n and T^2 modules Ω_+, Ω_-, N and M are defined as follows. Choose relatively prime positive integers p and q and n integers ω_i, $i = 1,2 \ldots n$. Define $\alpha = \frac{pq+1}{2}$ and $\beta = \frac{pq-1}{2}$. Then $|\Omega_+| = |\Omega_-| = \mathbb{H}^n$ and $|N| = |M| = \mathbb{H}$ where $\mathbb{H}^n$ and $\mathbb{H}$ are viewed as right complex vector spaces and as T^2 modules

(i) For $u \in \Omega_+$, $(\eta,t)u = u'$ $u'_i = \eta t^{\omega_i+\alpha} u_i t^{-\alpha}$, $i = 1,\ldots, n$.

(ii) For $v \in \Omega_-$, $(\eta,t)v = v'$ $v'_i = \eta t^{\omega_i+\alpha} v_i t^{\beta}$, $i = 1, \ldots n$.

(iii) For $z \in N$ $(\eta,t)z = t^{\frac{p+q}{2}} z t^{\frac{p-q}{2}}$

(iv) For $w \in M$ $(\eta,t)w = t^{\alpha} w t^{\beta}$.

As Sp_n modules, $\Omega_+ = \Omega_-$ is the standard Sp_n module with $|\Omega_+| = \mathbb{H}^n$ while $N = M$ is the trivial Sp_n module with $|N| = \mathbb{H}$.

Let $\Omega = \Omega_+ \oplus \Omega_-$. Then $\Omega \oplus N$ is a complex Sp_n module as well as a T^2 module; hence, defines homomorphisms of these two groups to U_{4n+2}. Let $G \subset U_{4n+2}$ be the subgroup generated by the images of T^2 and Sp_n under the above homomorphisms. Then Ω_+, Ω_-, N, M and Ω are complex G modules.

Choose positive integers a and b which satisfy $ap - bq = -1$. The map $\omega : N \to M$ defined by

$$\omega(z_0 + jz_1) = (z_0^q + \bar{z}_1^p) + jz_0^a z_1^b$$

is G equivariant, proper and $\dot{\omega}^{-1}(0) = 0$.
Here $z = z_0 + jz_1 \in \mathbb{H}$ and the z_i are the complex coordinates of z. Since $z \to |\omega(z)|^2$ is a real algebraic map, it can have only finitely many critical values ([10] p.16); so by multiplication by a suitable constant, we can suppose that this function has no critical values in $(0,1]$.

By restricting to $S^1 = T_2 \subset G$, we view N and M as S^1 modules. Then $N = t^p + t^{-q}$ and $M = t^{pq} + t^{-1}$.

Proposition 8.1 <u>The degree of</u> $|\omega| : \mathbb{H} \to \mathbb{H}$ <u>is</u> 1.

Proof: Since $\omega : N \to M$ is S^1 equivariant and proper, $f(t) = \lambda_{-1}(M)/\lambda_{-1}(N) \in R(S^1)$ and degree $|\omega| = f(1) = 1$.

The subgroup $T_1 \subset T^2$ acts freely on the sphere $S(\Omega)$. There is an induced $T_2 = S^1$ action on $P(\Omega) = S(\Omega)/T_1$. The S^1 manifold $P(\Omega)$ can also be described as the space of complex lines in the $S^1 = T_2$ module Ω.
Define a proper G map $F : S(\Omega) \times N \to M$ by $F(u,v,z) = 2 < u, v> - \omega(z)$. Then F induces a $T_2 = S^1$ map $\bar{F}$ from $P(\Omega) \times N$ to M. Set

$$Z(\omega) = F^{-1}(0)$$

$$X(\omega) = \bar{F}^{-1}(0) = Z(\omega)/T_1 .$$

Then $Z(\omega)$ is a G manifold and $X(\omega)$ is an S^1 manifold.

Proposition 8.2 $Z(\omega)$ <u>is a smooth</u> G <u>manifold</u>, $X(\omega)$ <u>is a smooth</u> S^1 <u>manifold and the</u> S^1 <u>normal bundle of</u> $X(\omega)$ <u>in</u> $P(\Omega) \times N$ <u>is</u> $X(\omega) \times M$.

roof: F and hence $\bar{F}$ is transverse regular to 0.
he essential point is that the function $f : \mathbb{H}^n \times \mathbb{H}^n \to \mathbb{H} \times R$ efined by $f(u,v) = (2\langle u,v\rangle, |u|^2 - |v|^2)$ has maximal rank t points other than $(0,0)$ [10] p.103 and the function $\to |\omega(z)|^2$ has no critical values in $(0,1]$.

heorem 8.3 <u>$|Z(\omega)|$ is homeomorphic to the unit sphere</u> <u>$(\mathbb{C}^{4n})$ and $|X(\omega)|$ is homeomorphic to $P(\mathbb{C}^{4n})$ the space</u> <u>f complex lines in $\mathbb{C}^{4n}$</u>.

roof: The circle $T_1 \subset T^2$ acts on $|Z(\omega)|$ via the estriction of an Sp_n action (8(i) and (ii)). The Sp_n omeomorphism of $|Z(\omega)|$ with $S(\mathbb{H}^n \times \mathbb{H}^n)$ of Theorem 7.5 arries the T_1 action on $|Z(\omega)|$ to a free linear action n $S(\mathbb{H}^n \times \mathbb{H}^n)$ and the quotient of this sphere by a ree linear action of $S^1 = T_1$ is clearly $P(\mathbb{C}^{4n})$. Thus $X(\omega)|$ is homeomorphic to $P(\mathbb{C}^{4n})$.

roposition 8.4 <u>The fixed point sets $X(\omega)^{S^1}$ and $P(\Omega)^{S^1}$</u> <u>re the same.</u>

roof: $(P(\Omega) \times N)^{S^1} = (P(\Omega) \times M)^{S^1} = P(\Omega)^{S^1}$ and $X(\omega) = \bar{F}^{-1}(0)$.

orollary 8.5 <u>Let $p \in X(\omega)^{S^1} = P(\Omega)^{S^1}$. Then the real S^1</u> <u>odules $TX(\omega)_p \oplus M$ and $TP(\Omega)_p \oplus N$ are equal.</u>

roof: This is immediate from (8.2).

In view of (8.5), we see that the construction of the ^{1} manifold $X(\omega)$ from $P(\Omega)$ has the effect of removing the S^1 module M as a factor of $TP(\Omega)_p$ and replacing it by the S^1 module N for each $p \in X(\omega)^{S^1}$.

Let $\tilde{\omega} : Z(\omega) \to S(\Omega)$ be the composition $Z(\omega) \subset S(\Omega) \times N \to S(\Omega)$ and let $\overline{\omega} : X(\omega) \to P(\Omega)$ be the map induced by $\tilde{\omega}$.

Theorem 8.7 $[X(\omega),\overline{w}] \in S_{S^1}(P(\Omega))$ <u>and</u> $[X(\omega), \overline{\omega}] \neq [P(\Omega), I_{P(\Omega)}]$; <u>moreover</u>, $\overline{\omega}_*(1_{X(\omega)}) = \lambda_{-1}(M)/\lambda_{-1}(N)\cdot 1_{P(\Omega)}$ <u>in</u> $K^0_{S^1}(P(\Omega))$.

Proof: Since θ is a homotopy equivalence (proof of (7.5)), it follows easily that $|\tilde{\omega}|$ is a homotopy equivalence which implies $|\overline{\omega}|$ is a homotopy equivalence. By (8.4) $|\overline{\omega}^{S^1}|$ is a homeomorphism of fixed point sets. This verifies the first assertion. The third assertion is a consequence of (5.4). The second follows from the third and (5.3) because $\overline{\omega}_*(1_{X(\omega)})$ is not a unit of $\tilde{K}^0_{S^1}(P(\Omega))$. Alternatively it follows from (6.4) and (8.5).

Remarks: (i) $\lambda_{-1}(M)/\lambda_{-1}(N) = \dfrac{(1-t^{pq})(1-t^{-1})}{(1-t^p)(1-t^{-q})}$ is divisible by $\phi_{pq}(t) \in R(S^1)$ so the torsion $\overline{\omega}_*(1_{X(\omega)})$ is not a unit of $\tilde{K}^0_{S^1}(P(\Omega))_\rho$, $\rho = (\phi_{pq}(t))$. This shows the set P_1 of (5.3) cannot be enlarged.

(ii) From (8.7) $\phi_{P(\Omega)}\overline{\omega}_*(1_{X(\omega)}) = 1$; so $|\overline{\omega}|^*$ preserves Pontrjagin classes by (6.2).

(iii) If the $4n$ integers $\{\omega_i+k \mid i=1,2\ldots n,\ k=0,1,pq,pq+1\}$ are distinct, $X(\omega)^{S^1}$ and $P(\Omega)^{S^1}$ consist of isolated points and the collection of representations $\{TX(\omega)_p \mid p \in X(\omega)^{S^1}\}$ is distinct from the collection $\{TP(\Omega')_q \mid q \in P(\Omega')^{S^1}\}$

or any S^1 module Ω' .

(iv) There is a simple description of the algebra $K^*_{S^1}(X(\omega))$. In particular it is a free $R(S^1)$ module of rank 4n; moreover, the cokernel of $\bar{\omega}^* : K^*_{S^1}(P(\Omega)) \to K^*_{S^1}(X(\omega))$ is the direct sum of 2n copies of $R(S^1)/\rho$ $\rho = (\phi_{pq}(t))$. We know from the first remark and (5.3), that $\bar{\omega}^*_\rho$ could not be an isomorphism.

(v) By comparing the fixed sets $X(\omega)^{Z_m}$ and $P(\Omega)^{Z_m}$ for $m = pq$, we see that the hypothesis that m be a prime power cannot be removed from (3.4) and from the work of [8] .

9. The case of finite isotropy groups.

In this section we indicate some applications of the materi of section I.4 to the study of S^1 manifolds X with finite isotropy groups. In order to keep the notation within bounds, we merely sketch the ideas involved.

Our setting is now the following: X and Y are closed S^1 manifolds and $f:X \to Y$ is an S^1 normal map i.e. f is equivariant and there is an S^1 bundle ξ over Y and a specifi S^1 bundle map $F:\nu_X \to \xi$ covering f. Here ν_X is the S^1 normal bundle of an imbedding of X in some real S^1 module. The bundle map F is part of the data of the normal map but wil not concern us here. The notion of S^1 normal cobordism is obvious. We suppose $Y^{S^1} = \emptyset$ and consider the following questions:

(a) How can we construct S^1 normal maps?

(b) Given an S^1 normal map $f:X \to Y$, when is this S^1 normally cobordant to $f':X' \to Y$ such that $|f'|$ is a homotopy equivalence?

(c) Given an S^1 normal map $f:X \to Y$ such that $|f|$ is a homotopy equivalence, what are the relations between the lattice of isotropy groups of X and Y? If K is an isotropy group of the action on X, what are the relations between the K modules $\nu(X^K,X)_x$ and

$\nu(Y^K, Y)_{f(x)}$ $x \in X^K$ defined by the representations of K on the normal fibers to X^K at x and Y^K at $f(x)$?

(d) How can we construct S^1 normal maps f such that $|f|$ is a homotopy equivalence and such that the lattice of isotropy groups of X is arbitrary subject to the relations of (c) and the set of representations $\nu(X^K, X)_x$ $x \in X^K$ is arbitrary (for K an isotropy group) subject to the relations of (c).

In this generality these questions are far from solved but here are some remarks that illuminates their motivation and their treatment in the context of these notes:

<u>Remarks</u>: (R. b) This is an S^1 cobordism question. We seek a complete set of numerical invariants of the S^1 action on X, Y (and f) such that if $f_i : X_i \longrightarrow Y$ have the same numerical invariants for $i = 1,2$, they are S^1 normally cobordant. This is a question of injectivity; namely to find an abelian group A and an injection of the set of S^1 normal maps to Y into A. A closely related problem is the question of when X_1 and X_2 are S^1 cobordant through a cobordism W with $W^{S^1} = \emptyset$. This question is easier to treat because of the group structure on the cobordism classes of S^1 manifolds X with $X^{S^1} = \emptyset$. It is clear that this problem should be treated before the more difficult problem of S^1 normal cobordism. In this context we

pose Question (9.20) below. This is the injectivity question for S^1 cobordism with finite isotropy groups.

(R. c) and (R. d). In order to illustrate these questions, we further restrict Y; namely we require that Y be a pseudo-free S^1 manifold in the sense of Montgomery-Yang. Thus we have $Y^{S^1} = \emptyset$ and if $y \in Y$ has isotropy group G_y different from 1, Y^{G_y} is a finite union of orbits. We remark that if $f: X \to Y$ is an S^1 normal map and Y is a pseudo-free S^1 manifold, so is X.

Suppose $f: X \to Y$ is an S^1 normal map with $|f|$ a homotopy equivalence and Y pseudo-free then. Let $\bar{x}$ denote the orbit of $x \in X$ and $G_{\bar{x}} = G_{x'}$ for any $x' \in \bar{x}$. Let $\bar{f}$ denote the orb map.

Theorem 1. The isotropy groups of X and Y are related by

$$G_{\bar{y}} = \prod_{\bar{x} \in \bar{f}^{-1}(\bar{y})} G_{\bar{x}} \qquad y \in Y \text{ and } x \in X.$$

Theorem 2. Let $x \in X$ and $\nu(X^{G_x}, X)_x = t^{a_1} + \ldots + t^{a_n}$ $\nu(Y^{G_{f(x)}}, Y)_{f(x)} = t^{b_1} + \ldots + t^{b_n}$ denote the indicated complex G_x modules. (We can suppose $\nu(X^{G_x}, X)$ and $\nu(Y^{G_y}, Y)$ are complex S^1 bundles if X and Y are oriented. In particular $\dim X = \dim Y = 2n+1$). Then

$$\pm \prod a_i \equiv \frac{|G_{f(x)}|}{|G_x|} \prod b_i \mod |G_x|.$$

ere $|G_x|$ *is the order of* G_x.

Theorem 3. *Let* $Y = S(V)$ *where* V *is the* S^1 *module* $t^q + nt^1$ *and* n *is odd. Given any splitting of the unique* *sotropy group* Z_q *of the* S^1 *action on* Y

$$Z_q = Z_{q_1} \times Z_{q_2} \times \cdots \times Z_{q_r}$$

nd given any Z_{q_i} *modules* V_i *with*

$$V_i = t^{a_{i1}} + \cdots + t^{a_{in}}$$

ith

$$\pm \prod_{j=1}^{n} a_{ij} \equiv q/q_i \bmod(q_i),$$

here is an S^1 *normal map* $f: X \longrightarrow Y$ *with* $|f|$ *a homotopy* *quivalence and there are exactly* r *isotropy groups of the* *ction on* X *which are* $Z_{q_1}, Z_{q_2}, \ldots, Z_{q_r}$ *and if* $x \in X$ *with* $G_x = Z_{q_i}$,

$$\nu(X^{G_x}, X)_x = V_i.$$

Remarks: For $n = 3$, Theorem 3 is due to Montgomery-Yang. strongly recomend their fundamental paper *Differentiable* *seudo-Free Circle Actions*, Proc. Nat. Acad. Sci. U.S.A., 894-896 1971). Actually Theorem 3 holds in a more general setting but he statement becomes more complicated.

The techniques used to prove the above theorems involved the ideas from equivariant K theory treated above; also the cobordism question (b) for pseudo-free S^1 actions had to be solved. Many problems of classification of actions with finite isotropy groups remain in particular for pseudo-free actions and it is to be expected that the techniques and invariants of these lecture notes will come to play again especially $f_*(1_X) \in \tilde{K}_{S^1}(Y)$ when $f: X \longrightarrow Y$ is an S^1 normal map.

Here are two questions which appear useful and interesting.

(e) Determine $S_{S^1}(Y)$ when $Y^{S^1} = \phi$. E.g. when $Y = S(M)$ and M is a complex S^1 module with $M^{S^1} = \phi$.

(f) If $[X,f] \in S_{S^1}(Y)$, $Y^{S^1} = \phi$, what can be said about $K^*_{S^1}(X)$. (When $Y = S(t^q + nt^1)$ Janey Daccach has computed $K^*_{S^1}(X)$ and found $f^*: K^*_{S^1}(Y) \to K^*_{S^1}(X)$ is surjective.)

(R. a) In order to construct S^1 normal maps $f: X \longrightarrow Y$, we must treat the question of equivariant transversality and S^1 quasi-equivalent bundles over Y. Suppose ξ and η are S^1 vector bundles over Y. We say that ξ and η are <u>quasi-equivalent</u> if there exists a proper S^1 map $\omega: \xi \to \eta$ which preserves fibers and $|\omega|: |\xi_y| \longrightarrow |\eta_y|$ has degree 1 for each $y \in Y$. (This is not an equivalence relation as there may be no such map from η to ξ.) Given a quasi-equivalence $\omega: \xi \to \eta$, when is ω properly S^1 homotopic to a map $\theta: \xi \longrightarrow \eta$ such hat Θ is transverse to the zero section $Y \subset \eta$ written $\Theta \pitchfork Y$? If $\Theta \pitchfork Y$ then $X = \Theta^{-1}(Y)$ is an S^1 manifold and $\Theta|_X = f: X \longrightarrow Y$ is an S^1 normal map.

Thus to answer (a), we must classify quasi-equivalent S^1 ector bundles over Y and solve the transversality problem or quasi-equivalences. Both are hard problems. When Y is a oint, Alan Meyerhoff in his thesis, has classified the quasi-quivalent vector bundles over Y. This has many applications nd is implicit in the three theorems mentioned above. The ransversality problem is in a certain sense solved in Petrie, _bstructions to transversality for compact Lie Groups_, to appear n Bull. A.M.S. and will be treated more in detail in a latter aper.

The above three theorems and many other ideas on S^1 anifolds with finite isotropy groups will appear in latter papers.

The remainder of this section is devoted to the S^1 obordism problem (b).

Let $\mathfrak{F}$ denote a family of subgroups of S^1 which is either o = 1 or $\mathfrak{F}_1$ the family of all finite subgroups of S^1. Denote y $\Omega_i^{S^1}(\mathfrak{F})$ the bordism group of oriented closed i dimensional 1 manifolds whose isotropy groups are in the family $\mathfrak{F}$. If X s such a manifold, its class in $\Omega_i^{S^1}(\mathfrak{F})$ is denoted by $[X]$. The roups $\Omega_*^{S^1}(\mathfrak{F}) = \Sigma_i \Omega_i^{S^1}(\mathfrak{F})$, are modules over $\Omega_*(pt)$, the oriented bordism ring of Thom. What is the structure of these groups?

There is an isomorphism (of Ω_* (pt) modules).

(9.1) $$F:\Omega_*^{S^1}(\mathfrak{F}_0) \longrightarrow \Omega_*(B_{S^1})$$

defined by $F[X] = [X/S^1, f_X]$ where $\Omega_*(B_{S^1})$ is the bordism of the classifying space B_{S^1} of S^1 and $f_X: X/S^1 \longrightarrow B_{S^1}$ is a map which classifies the principle S^1 bundle $X \longrightarrow X/S^1$. (See Conner-Floyd- , Differentiable Period Maps, Springer, 1964.) This gives the structure of $\Omega_*^{S^1}(\mathfrak{F}_0)$.

In order to discuss $\Omega_*^{S^1}(\mathfrak{F}_1)$ we relate the material of section 3 with Atiyah[+]. It is convenient to complexify all groups under discussion by tensoring with the complex numbers C. We assume this done and do not explicitly indicate it in the notation.

Denote the C^∞ complex valued functions on S^1 by $D(S^1)$ and its (topological) dual vector space $\mathrm{Hom}_C^t(D(S^1), C)$ by $D'(S^1)$ the space of distributions on S^1. The latter is a module over $R(S^1)$ by defining a representation to act by multiplication by its character. There is a homomorphism

[+]Elliptic Operators and Compact Groups, Lecture 9, Springer Verla (1974).

9.2) $$J: F/R \longrightarrow D'(S^1)$$

R = $R(S^1) \otimes_{\mathbb{Z}} \mathbb{C}$ and F is the field of fractions of R)

efined by

9.3) $$J(f)(g) = \sum_{|t|=1} \text{residue } \frac{f(t)g(t)}{t}$$

or $f(t)$ a rational function of $t \in S^1$ representing an

lement of F/R and $g = g(t)$ an element of $D(S^1)$. We denote

ocalization at the prime ideal $(t-1)$ of R with a subscript

. Then

9.4) $$(F/R)_1 \cong C[t,t^{-1}](\tfrac{1}{1-t}) / C[t,t^{-1}]$$

s the (topological) dual of $C[[t-1]]$ the ring of power series

n t-1. This ring is $R(S^1) \hat{\otimes} C$ the completion of $R(S^1) \otimes C$

t the ideal (t-1). It is isomorphic to $K^0(B_{S^1}) \hat{\otimes} C$. The

ocalization $D'(S^1)_1$ is the space of distributions supported

y 1 so is the (topological) dual of $C[[t-1]]$ i.e.

9.5) $$D'(S^1)_1 \cong \text{Hom}^t_C(C[[t-1]],C).$$

The localization J_1 of J induces an isomorphism

(9.6) $$J_1 : (F/R)_1 \longrightarrow \mathrm{Hom}^t_C(C[[t-1]], C)$$

$$J_1(f)(g) = \operatorname*{Residue}_{t=1} \frac{f(t)g(t)}{t}$$

Observe that an element of $(F/R)_1$ is represented by an expansio[n]

(9.7) $$\sum_{i=-1}^{-N} a_i (t-1)^i, \quad a_i \in C \quad \text{for some integer } N.$$

Let $SO(2\ell)$ denote the orientation preserving isometries of $R^{2\ell}$. The cohomology of its classifying space $H^*(B_{SO(2\ell)})$ is a subring of $C[x_1,\ldots,x_\ell]$ which contains the ring of symmetric polynomials in the x_i^2. The ith elementary symmetric function of the $\{x_j^2\}$ is the ith Pontrjagin class $p_i \in H^{4i}(B_{SO(2\ell)})$

Let $H^{**}(B_{SO(2\ell)}) = \prod_{i=0}^{\infty} H^i(B_{SO(2\ell)})$ and let $L^{-1}(p_1,\ldots,p_\ell) \in H^{**}(B_{SO(2\ell)})$ denote the inverse of the Hirzebruch class and let $\sigma = \sigma(t_1,\ldots,t_\ell)$ be any symmetric function of degree $<2\ell$.

here is a $u_\sigma \in R(SO(2\ell))$ such that

$$(9.8) \qquad ch' u_\sigma = L^{-1}(p_1,\dots,p_\ell)\sigma(x_1^2,\dots,x_\ell^2) + \omega \in H^{**}(B_{SO(2\ell)})$$

here $\omega \in \prod_{i>2\ell} H^{**}(B_{SO(2\ell)})$ and ch' is the composition

$$R(SO(2\ell)) \longrightarrow K^0(B_{SO(2\ell)}) \xrightarrow{ch} H^{**}(B_{SO(2\ell)}).$$

ee [2] p. 596.

Let V be an oriented G bundle of even dimension over a manifold M and let $\alpha(V) \in K_G^0(V)$ be its index class [2] . 576. In particular for $G = 1$ and M an oriented 2ℓ imensional manifold $\alpha(TM) \in K^0(TM)$ and if $u_\sigma(TM) \in \hat{K^0}(M)$ is he element associated to TM by u_σ, we have for any $\beta \in K^0(M)$,

$$(9.9) \qquad Id^M(\alpha(TM)u_\sigma(TM)\beta) =$$

$$= \langle ch u_\sigma(TM) ch(\beta) L(TM), [M]\rangle \quad [2] \text{ p. } 586$$

$$= \langle \sigma(TM) ch(\beta), [M]\rangle \quad \text{by } (9.8).$$

ere $L(TM) \in H^*(M)$ is the Hirzebruch class of TM and $\sigma(TM)$ s the polynomial in the Pontrjagin classes of TM defined by σ.

Let Σ denote the graded ring $\mathbb{C}[\sigma_1,\sigma_2,\dots]$ where degree $_i = 4i$ and let $Hom^*_{\mathbb{C}}(\Sigma,A)$ be the direct sum of the $Hom_{\mathbb{C}}(\Sigma_d,A)$ here $\Sigma_d \subset \Sigma$ is the part of degree d. Then we have a homo- orphism

(9.10) $$\psi:\Omega_*(B_{S^1}) \longrightarrow \mathrm{Hom}^*(\Sigma, D'(S^1)_1)$$

defined for $[M,g] \in \Omega_{2\ell}(B_{S^1})$ by

$$\psi[M,g](\theta)(\eta) = \mathrm{Id}^M(\alpha(TM)\, u_{\theta_\ell}(TM)\, g^*(\eta))$$

for $\theta \in \Sigma$ and $\eta \in K^0(B_{S^1}) \hat{\otimes} C \cong \mathbb{C}[[t-1]]$. Here θ_ℓ is the symmetric function of ℓ variables defined as the image of θ under the algebra homomorphism from Σ to the ring of symmetric functions in $\{x_1^2, \ldots, x^2\}$ which sends σ_i to the ith elementary symmetric function of the x_j^2 for $i \leq \ell$, and sends σ_i to zero for $i > \ell$. Since $K^0(M) \cong K^0(\hat{M})$, $g^*(\eta)$ makes sense.

(9.11) Remark: ψ is an isomorphism. This follows from the fact that $ch: K^0(B_{S^1}) \hat{\otimes} \mathbb{C} \longrightarrow H^{**}(B_{S^1})$ is an isomorphism, $D'(S^1)_1 = \mathrm{Hom}^t_{\mathbb{C}}(K^0(B_{S^1}) \otimes \mathbb{C}, \mathbb{C})$, the Atiyah-Hirzebruch spectral sequence collapses $H_*(B_{S^1}, \Omega_*(pt)) \Rightarrow \Omega_*(B_{S^1})$, and elementary calculations with Pontrjagin numbers using (9.9).

Let X be a closed $2\ell+1$ dimensional manifold with $X^{S^1} = \emptyset$. Let $TO \subset TX$ be the subbundle of vectors tangent to the orbits. i.e. for $x \in X$, TO_x is the tangent space of the orbit $S^1 x$. TO is the one dimensional trivial bundle $\epsilon^1 = X \times R^1$. There is an exact sequence

(9.13) $$0 \longrightarrow TO \longrightarrow TX \longrightarrow T_{S^1}X \longrightarrow 0$$

which defines $T_{S^1}X$ as an S^1 vector bundle over X. If X is oriented, so is $T_{S^1}X$ and

(9.14) $$TX = T_{S^1}X \oplus \epsilon^1 .$$

Atiyah in Elliptic Operators and Compact Groups defines

(9.15) $$\mathrm{id}^X : K^0_{S^1}(T_{S^1}X) \longrightarrow D'(S^1)$$

which is related to $\mathrm{Id}^X_{F/R}$ via

(9.16) $$\begin{array}{ccc} h_1(X) = K^1_{S^1}(TX) = & & K^0_{S^1}(T_{S^1}X) \\ \big\downarrow \mathrm{Id}^X_{F/R} & & \big\downarrow \mathrm{id}^X \\ F/R & \xrightarrow{\ J\ } & D'(S^1) \end{array}$$

Moreover he computes explicitly id^X in the case S^1 acts freely on X. In that case $K^0_{S^1}(T_{S^1}X) = K^0_{S^1}(T_{S^1}X)_1$, $T_{S^1}X/S^1 =$ TX/S^1 and $\mathrm{id}^X = \mathrm{id}^X_1$ (the localization of id^X at $(t-1)$). There is a commutative diagram

(9.17) $$\begin{array}{ccc} K^0_{S^1}(T_{S^1}X) & \xrightarrow{\ \mathrm{id}^X_1\ } & D'(G)_1 \\ \big\uparrow P^* & \nearrow \mu & \\ K^0(TX/S^1) & & \end{array}$$

where P^* is the isomorphism arising from $T_{S^1}X/S^1 = TX/S^1$

and $\mu(\alpha)(\eta) = Id^{X/S^1}(\alpha \cdot f_X^*(\eta))$

$$\alpha \in K^0(TX/S^1), \eta \in K^0(B_{S^1}), \ f_X : X/S^1 \longrightarrow B_{S^1}$$

the map classifying $X \longrightarrow X/S^1$.

Let X be an odd dimensional S^1 manifold with finite isotropy groups. Then $h_1(X) = K^1_{S^1}(TX) = K^0_{S^1}(TX \oplus \varepsilon^1)$ where ε^1 is the real one dimensional trivial bundle over X. Since $TX \oplus \varepsilon^1$ is even dimensional, $\alpha(TX \oplus \varepsilon^1) \in K^1_{S^1}(TX) = K^0_{S^1}(TX \oplus \varepsilon^1)$ and we define $\beta_X \in h_1(X)$ by

$$\beta_X = \alpha(TX \oplus \varepsilon^1) \tag{9.18}$$

We define a homomorphism

$$\Omega_*^{S^1}(\mathfrak{F}_1) \xrightarrow{\ \ominus\ } \mathrm{Hom}_{\mathbf{C}}^*(\Sigma, F/R) \tag{9.19}$$

by

$$\ominus[X](\theta)(\eta) = J[Id^X_{F/R}(\beta_X \cdot u_{\theta_\ell}(T_{S^1}X)](\eta)$$

if

$$[X] \in \Omega^{S^1}_{2\ell+1}(\mathfrak{F}_1), \theta \in \Sigma, \eta \in D(S^1)$$

and $$\ominus[X] = 0 \quad \text{if} \quad \Omega^{S^1}_{2\ell}(\mathfrak{F}_1).$$

Remarks: $J: F/R \longrightarrow D'(S^1)$ is a monomorphism so Θ is ell defined. $u_{\theta_\ell}(T_{S^1}X) \in K^0_{S^1}(X)$ is associated to $T_{S^1}X$ by $\theta_\ell \in R(SO(2\ell))$. Observe that the structure group of $|T_{S^1}X|$ s $SO(2\ell)$.

(9.20) Question. Is Θ a monomorphism and if not what is er Θ ?

If $[X] \in \Omega_*^{S^1}(\mathfrak{J}_0)$, it is easy to see that $\Theta[X] \in \mathrm{Hom}^*_{\mathbb{C}}(\Sigma,(F/R)_1)$ and we call the induced homomorphism Θ_1

9.21) $$\Theta_1 : \Omega_*^{S^1}(\mathfrak{J}_0) \longrightarrow \mathrm{Hom}^*_{\mathbb{C}}(\Sigma,(F/R)_1).$$

Proposition 9.22. Θ_1 <u>is an isomorphism</u>.

Proof: The following diagram is commutative:

$$\begin{array}{ccc}
\Omega_*^{S^1}(\mathfrak{J}_0) & \xrightarrow{\ \Theta_1\ } & \mathrm{Hom}^*_{\mathbb{C}}(\Sigma,(F/R)_1) \\
\downarrow F & & \downarrow \mathrm{Hom}^*_{\mathbb{C}}(\Sigma, J_1) \\
\Omega_*(B_{S^1}) & \xrightarrow{\ \psi\ } & \mathrm{Hom}^*_{\mathbb{C}}(\Sigma, D'(S^1)_1)
\end{array}$$

his follows from (9.16) and (9.17). As previously remarked , ψ and J_1 are isomorphisms hence Θ_1 is an isomorphism.

(9.23) Does $\Omega_*^{S^1}(\mathfrak{J}_1)$ have the structure of an R module n such a way that $\Omega_*^{S^1}(\mathfrak{J}_0) = \Omega_*^{S^1}(\mathfrak{J}_1)_1$?

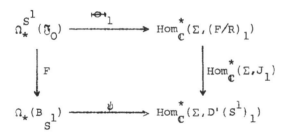

(9.24) Remark: The point of Proposition 9.22 is that it motivates Question (9.14). Ossa (Math. Ann. 186(1970), 45-52) has shown that the groups $\Omega_*^{S^1}(\mathfrak{F}_1)$ are not finitely generated over $\Omega_*(pt)$. Perhaps some additional algebraic structure (such as an R module structure on $\Omega_*^{S^1}(\mathfrak{F}_1)$) would give more understanding to these groups.

Two other potential applications of section 3 are (a) study of the set $S_{S^1}(S(M))$ where M is a complex S^1 module with $M^{S^1} = \phi$ (b) use of the invariant $\sigma_X = J(Id^X_{F/R}(\beta_X))(1)$, $1 \in D(S^1)$ to obtain surgery and transversality obstructions. If $X^{S^1} = \phi$, then X/S^1 is a rational homology manifold, so has a signature $I(X/S^1)$. Atiyah has shown that

$$I(X/S^1) = \sigma_X$$

and given an explicit formula for σ_X in terms of the invariants of the S^1 action on X.

As an application consider the following situation. Suppose X and Y are S^1 manifolds of the same dimension and $X^{S^1} = Y^{S^1} = \phi$. In addition suppose that $f: X \longrightarrow Y$ is an S^1 map. If $|f|: |X| \longrightarrow |Y|$ is a homotopy equivalence, $I(X/S^1) = I(Y/S^1)$. Since $I(X/S^1)$ is an $\Omega^{S^1}(\mathfrak{F}_1)$ cobordism invariant, we see that $I(X/S^1) - I(Y/S^1)$ is an obstruction to finding an S^1 cobordism W with $W^{S^1} = \phi$, $\partial W = X \cup X'$ and an S^1 map

$':X' \longrightarrow Y$ such that $|f'|$ is a homotopy equivalence.

If N and M are S^1 vector bundles over Y and $':N \longrightarrow M$ is a proper S^1 map, when is F' properly S^1 omotopic to F with F transverse to $Y \subset M$ in such a way hat F restricted to $F^{-1}(Y) = X$ is a homotopy equivalence? here are two problems involved (a) transversality (b) making $|_X$ a homotopy equivalence. As an application of the formula f Atiyah for the signature $I(Y/S^1)$ in terms of geometric ata, there is a rational number $\sigma(F')$ only depending upon he proper S^1 homotopy class of F' such that (a') $\sigma(F')$ is n integer if F' is properly S^1 homotopic to a transversal ap F. (b') $\sigma(F') = 0$ if $F:F^{-1}(X) \longrightarrow F^{-1}(Y)$ is a homotopy quivalence.

These remarks have obvious application to the study of $_{S^1}(Y)$ when $Y^{S^1} = \emptyset$. Observe that $f_*(1_X)$ is defined when $X,f] \in S_{S^1}(Y)$. This invariant should have important applications n case $Y^{S^1} = \emptyset$ just as in the case $Y^{S^1} \neq \emptyset$ which was studied n detail earlier.

10. Induction

This section presents some algebraic situations and problem which arise form comparing two R valued bilinear forms $\langle\ \rangle_\Lambda$ $\langle\ \rangle_\Gamma$ on R orders Λ and Γ. Here R will be a P.I.D. which arises by localizing $R(S^1)$, the complex representation ring of S^1, at a set P of prime ideals in $R(S^1)$. These forms occur geometrically in the following way: Let X and Y be closed smooth S^1 manifolds of even dimension. Under mild assumptions on X and Y, there are non-degenerate $R = R(S^1)_P$ valued symmetric bilinear forms on $\Lambda = \tilde{K}^*_{S^1}(X)_P$ and $\Gamma = \tilde{K}^*_{S^1}(Y)_P$ constructed from the algebraic structure of each and the Atiyah-Singer Index homomorphism [2]. $\tilde{K}^*_{S^1}(\ )_P$ is $K^*_{S^1}(\ )_P$ mod torsion (I,6

An S^1 map $f: X \longrightarrow Y$ induces an R algebraic homomorphis $f^*: \Lambda \longrightarrow \Gamma$. A natural geometric assumption concerning f leads to the situation in which f^* is a monomorphism and Λ, Γ, Θ = I are R orders in the semisimple $\bar{F}$ algebra $\Theta \otimes_R \bar{F}$. Using the non-degenerate bilinear forms $\langle\ \rangle_\Lambda$ and $\langle\ \rangle_\Gamma$ we define an induction homomorphism $f_*: \Gamma \longrightarrow \Lambda$. In particular, $f_*(1_X) \in \Lambda$ i an interesting algebraic invariant of the situation.

Knowledge of this invariant $f_*(1_X)$ translates into important geometric information comparing the differential struct of X and Y and the representations of S^1 on the normal bund to the fixed set $X^{S^1} \subset X$ and on the normal bundle to the fixed set $Y^{S^1} \subset Y$.

For now Γ will denote a commutative R algebra with identity 1, which is free of finite rank over R. Then there a wo cannonical homomorphisms from Γ to R

10.1) $\det_{\Gamma} : \Gamma \longrightarrow R$. Each element $\gamma \in \Gamma$ defines an R linear transformation from Γ to itself, namely left multiplication. Choosing an R base for Γ represents this transformation as a matrix over R. The determinant of this transformation is called $\det_{\Gamma}(\gamma)$. Thus $\det_{\Gamma}$ is a multiplicative homomorphism.

10.2) $tr_{\Gamma} : \Gamma \longrightarrow R$. As above, view an element $\gamma \in \Gamma$ as an R linear transformation. Its trace is called $tr_{\Gamma}(\gamma)$. Thus tr_{Γ} is an R linear homomorphism.

An R order in Γ denotes a subalgebra Λ of Γ such that $\otimes_R \bar{F} = \Gamma \otimes_R \bar{F}$. Let $f^*:\Lambda \longrightarrow \Gamma$ be the inclusion map. We shall ssume that Λ and Γ are equipped with R module homomorphisms $d^{\Lambda}:\Lambda \longrightarrow R$, $Id^{\Gamma}:\Gamma \longrightarrow R$ which induce non-degenerate bilinear forms. bilinear form is defined by $<\gamma_1, \gamma_2>_{\Gamma} = Id^{\Gamma}(\gamma_1 \cdot \gamma_2)$ for $\gamma_1, \gamma_2 \in \Gamma$. on-degenerate means that the R-homomorphism $\Phi(< , >_{\Gamma}) : \Gamma \longrightarrow Hom_R(\Gamma, R)$ efined by $\gamma \longmapsto <\gamma, >_{\Gamma}$ is an isomorphism.

With this in hand we can define the <u>induction homomorphism</u> $_*:\Gamma \longrightarrow \Lambda$ by

$$<f_*\gamma, \lambda>_{\Lambda} = <\gamma, f^*\lambda>_{\Gamma} \quad \text{for} \quad \gamma \in \Gamma, \quad \lambda \in \Lambda.$$

Using an abuse of notation we let $<\gamma_1, \gamma_2>_{\Gamma} = Id^{\Gamma}(\gamma_1 \cdot \gamma_2)$ ean $Id^{\Gamma} \otimes_R 1_{\bar{F}}(\gamma_1 \cdot \gamma_2)$ when $\gamma_1, \gamma_2 \in \Gamma \otimes_R \bar{F}$. $1_{\bar{F}}$ is the identity n $\bar{F}$. The bilinear form $< , >_{\Gamma}$ on $\Gamma \otimes_R \bar{F}$ is, of course, non-degenerate. Moreover, an element $\alpha \in \Gamma \otimes_R \bar{F}$ is completely determined by the R homomorphism $\phi:\Gamma \longrightarrow \bar{F}$ defined by $\gamma \longrightarrow \langle\alpha,\gamma\rangle_\Gamma$. In fact, since $\Lambda \otimes_R \bar{F} = \Gamma \otimes_R \bar{F}$, α is already determined by ϕ restricted to $f^*(\Lambda)$.

<u>Proposition 10.3</u> <u>Given</u> $\alpha_\Lambda \in \Lambda$, $\alpha_\Gamma \in \Gamma$, <u>which are units in</u> $\Lambda \otimes_R \bar{F} = \Gamma \otimes_R \bar{F}$, <u>then the following are equivalent</u>:

(i) $\mathrm{Id}^\Lambda(\lambda) = \mathrm{Id}^\Gamma(\frac{\alpha_\Gamma}{f^*\alpha_\Lambda} \cdot f^*\lambda)$ for all $\lambda \in \Lambda$

(ii) $f^*f_*(\gamma) = \frac{f^*\alpha_\Lambda}{\alpha_\Gamma} \cdot \gamma$ for all $\gamma \in \Gamma$

(iii) $f_*(\alpha_\Gamma) = \alpha_\Lambda$.

<u>Moreover, the above imply</u>

(iv) $f^*f_*(\gamma) = f^*f_*(1)\cdot\gamma$ for all $\gamma \in \Gamma$. 1 is the identit of Γ.

(v) $f_*(f^*(\lambda)\cdot\gamma) = \lambda\cdot f_*(\gamma)$ for all $\lambda \in \Lambda$, $\gamma \in \Gamma$.

<u>Proof</u>: We show (i) implies (ii):

$$\langle\frac{\alpha_\Gamma}{f^*\alpha_\Lambda}, f^*\lambda \cdot f^*f_*(\gamma)\rangle_\Gamma = \mathrm{Id}^\Gamma(\frac{\alpha_\Gamma}{f^*\alpha_\Lambda} \cdot f^*(f_*(\gamma)\cdot\lambda) = \mathrm{Id}^\Lambda(f_*(\gamma)\cdot\lambda)$$

$$= \langle f_*(\gamma),\lambda\rangle_\Lambda = \langle\gamma,f^*\lambda\rangle_\Gamma \quad \text{for all } \lambda \in \Lambda,$$

thus $\frac{\alpha_\Gamma}{f^*\alpha_\Lambda} \cdot f^*f_*(\gamma) = \gamma$ or $f^*f_*(\gamma) = \frac{f^*\alpha_\Lambda}{\alpha_\Gamma} \cdot \gamma$ for all $\gamma \in$

oposition 10.4: *Let* $\det_\Gamma : \Gamma \longrightarrow R$ *be the determinant homomorphism.* *en* $\det_\Gamma f^* f_*(1) \in U(R)\cdot R^2$ *where* $U(R)$ *denotes the multiplica-* *ve group of units of* R *and* R^2 *the subset of squares of* R.

oof: If we take an R base for Λ and for Γ, then f^* is presented by a square matrix over R. The homomorphism f_* is e composition of these homomorphisms:

$$\Gamma \xrightarrow[\approx]{\Phi(<\ >_\Gamma)} \mathrm{Hom}_R(\Gamma,R) \xrightarrow{f^{**}} \mathrm{Hom}_R(\Lambda,R) \xrightarrow[\approx]{\Phi(<\ >_\Lambda)^{-1}} \Lambda$$

ere f^{**} is $\mathrm{Hom}_R(f^*,R)$. If we take a dual base for Γ and a al base for Λ we obtain isomorphisms

$$j_\Gamma : \Gamma \longrightarrow \mathrm{Hom}_R(\Gamma,R) \quad \text{and} \quad j_\Lambda : \Lambda \longrightarrow \mathrm{Hom}_R(\Lambda,R)$$

d the matrix representing $j_\Lambda^{-1} f^{**} j_\Gamma$ is the transpose of f^*. ere are isomorphisms $P_\Gamma : \Gamma \longrightarrow \Gamma$ and $P_\Lambda : \Lambda \longrightarrow \Lambda$ such that

$$j_\Gamma P_\Gamma = \Phi(<\ >_\Gamma), \qquad j_\Lambda P_\Lambda = \Phi(<\ >_\Lambda).$$

us $f_* = \Phi(<\ >_\Lambda)^{-1} f^{**} \Phi(<\ >_\Gamma) = P_\Lambda^{-1} j_\Lambda^{-1} f^{**} j_\Gamma P_\Gamma$; so

$$\mathrm{t}f_* = (\det P_\Lambda)^{-1} \cdot \det(j_\Lambda^{-1} f^{**} j_\Gamma) \cdot \det(P_\Gamma) = (\det P_\Lambda)^{-1} \det P_\Gamma \det f^*$$

cause the determinant of the transpose of a matrix is the determi- nt of the matrix.

From Proposition 10.3 we have

$$f^*f_*(\gamma) = f^*f_*(1)\cdot\gamma \quad \text{for} \quad \gamma \in \Gamma.$$

This means that

$$\det f^* \det f_* = \det_\Gamma f^*f_*(1).$$

But $\det f^* \det f_* = (\det P_\Lambda)^{-1}\det P_\Gamma \cdot (\det f^*)^2$ from above. Since P_Λ and P_Γ are isomorphisms $(\det P_\Lambda)^{-1}$ and $\det P_\Gamma$ ar in $U(R)$.

<u>Corollary 10.5: The element</u> $\det_\Lambda (f_*(1))$, <u>is in</u> $U(R)\cdot R^2$.

Proof: This follows from the fact that $\det_\Gamma \circ f^* = \det_\Lambda$.

Remark: Actually we have proved a stronger statement which is useful to have. Here is an appealing way to state it: Since R is a principle ideal domain, any finitely generated R module M is a direct sum of cyclic modules, say

$$M = R/(P_1) \oplus \cdots \oplus R/(P_n)$$

where (P_i) denotes the principle ideal generated by $P_i \in R$.

<u>Definition</u>: The product ideal $(P_1\cdot P_2 \cdots P_n)$ is called the ord of M, written $\operatorname{ord}_R(M)$.

<u>Corollary 10.6</u>. <u>Let</u> A_1, A_2 <u>be two</u> R <u>algebras which are free</u>

modules of the same rank: Let $i: A_1 \longrightarrow A_2$ *be an inclusion of algebras.* Let $Id^{A_2} \in Hom_R(A_2, R)$, $a_1 \in A_1$, *and suppose the bilinear forms* $< >_{A_1}$ *and* $< >_{A_2}$ *defined by*

$$<x,y>_{A_1} = Id^{A_2} \otimes 1_F\left(\frac{i(x \cdot y)}{i(a_1)}\right), \qquad x,y \in A_1$$

$$<z,w>_{A_2} = Id^{A_2}(z \cdot w), \qquad z,w \in A_2$$

are R *valued and non-degenerate.* *Let* $i_*: A_2 \longrightarrow A_1$ *be the induction homomorphism.* *Then* $i_*(1) = a_1 \in A_1$ *and the* R *ideals* $(\det_{A_1} i_*(1))$ *and* $(\mathrm{ord}_R(A_2/A_1))^2$ *are equal.*

Suppose now we are given a commutative diagram

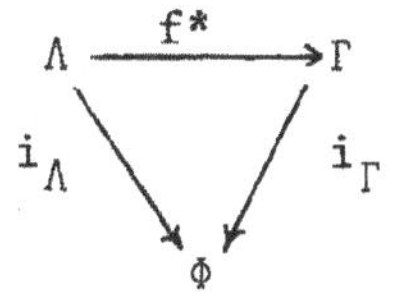

of inclusions of free commutative algebras such that Λ and Γ are orders in Φ. Let the R homomorphisms Id^Λ, Id^Γ, Id^Φ induce non-degenerate bilinear forms and assume that $(i_\Lambda)_*(1) = \alpha_\Lambda$, $(i_\Gamma)_*(1) = \alpha_\Gamma$ are units over F. It then follows that $f_*(\alpha_\Gamma) = \alpha_\Lambda$ and we have a commutative diagram of induction homomorphisms satisfying the properties of Proposition 10.3

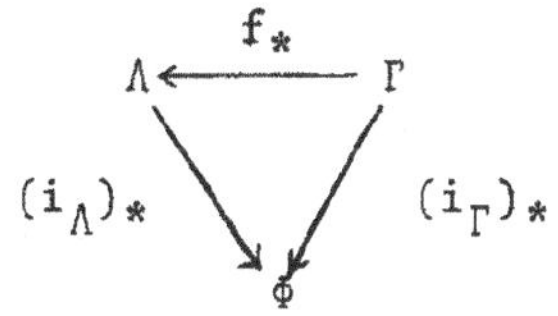

We further assume that if $\Theta = \Pi R$ is the product of copies of R then Id^{Θ} is the trace homomorphism tr_{Θ}.

Remark: More generally we want our algebras to be graded by integers mod 2. So $\Theta = \Theta_0 \oplus \Theta_1$ and $\Theta_i \Theta_j \subset \Theta_{i+j}$ where $i+j$ is taken mod 2. Moreover, if $\theta \in \Theta_i$ and $\theta' \in \Theta_j$ then $\theta \cdot \theta' = (-1)^{i \cdot j} \theta' \cdot \theta$. We reserve this remark for meditation after th contents of the simpler situation have been digested.

Corollary 10.7: The elements $f_*(1) \in \Lambda$, $\alpha_\Lambda = (i_\Lambda)_*(1) \in \Lambda$ and $\alpha_\Gamma = (i_\Gamma)_*(1) \in \Gamma$ are related by

$$i_\Lambda f_*(1) = i_\Lambda(\alpha_\Lambda) / i_\Gamma(\alpha_\Gamma) \in \Theta$$

Proof: We have from Proposition 10.3 (ii) that $f^*f_*(1) = \dfrac{f^*\alpha_\Lambda}{\alpha_\Gamma}$. Applying i_Γ gives $i_\Lambda f_*(1) = \dfrac{i_\Lambda \alpha_\Lambda}{i_\Gamma \alpha_\Gamma}$.

In light of the results of Section 3, we impose this additional condition on the orders Λ, Γ with $\Lambda \subset \Gamma$:

$f^*_{P_1} : \Lambda_{P_1} \longrightarrow \Gamma_{P_1}$ is an isomorphism, then

Corollary 10.8. $f_*(1)$ is a unit of Λ_{P_1}.

Proof: From the equation $f^*f_*(\gamma) = f^*f_*(1) \cdot \gamma$ and the fact that $f^*_{P_1}$ is an isomorphism (hence $(f_*)_{P_1}$ is an isomorphism), we conclude that $f^*f_*(1)$ is a unit of Γ_{P_1}. Since f^* is an algebra morphism, $f_*(1)$ is a unit of Λ_{P_1}.

Corollary 10.9. For any algebra homomorphism $r:\Theta \longrightarrow R$, $||rf_*(1)||_p = 1$ for all $p \in P_1$. See II §2.

Proof: $rf_*(1)$ is a unit of R_p.

An especially interesting case to study for applications occurs when $\Theta = \Pi_{i=1}^n R$ is the product of n copies of R. Here we can even enrich the structure already at hand with the following assumption: The element $\alpha_\Lambda \in \Lambda$ (also $\alpha_\Gamma \in \Gamma$) has this form: For some integer m, called the dimension of α_Λ, there are n complex representations of dimension m $\Lambda_1, \Lambda_2, \cdots \Lambda_n$ of S^1 (respectively $\Gamma_1, \Gamma_2, \cdots \Gamma_n$) and n integers μ_i (respectively v_i) such that

$$i_\Lambda(\alpha_\Lambda)_i = t^{\mu_i}\lambda_{-1}(\Lambda_i) \in R \qquad i = 1,2, \cdots n$$

$$i_\Gamma(\alpha_{\Gamma_i}) = t^{v_i}\lambda_{-1}(\Gamma_i) \in R \qquad i = 1,2, \cdots n$$

Here $\lambda_{-1}(M) = \sum(-1)^i\lambda^i(M)$ is the alternating sum of the exterior powers of the complex S^1 representation M. It is an element of $R(S^1) \subset R$.

There is an abundant set of algebra endomorphisms ψ^k $k = 0,1, \cdots n \cdots$ of R. These are the Adams operations. If $a/b \in R$ then

$$\psi^k(a/b) = a(t^k)/b(t^k) \in R$$

for $a = a(t)$, $b = b(t) \in R(S^1) = Z[t,t^{-1}]$.

These operations define algebra endomorphisms ψ^k of 0. Precisely if $\theta = (\theta_1, \theta_2, \cdots \theta_n) \in \Theta$ $\theta_i \in R$ $\psi^k(\theta) = (\psi^k(\theta_1), \cdots \psi^k(\theta_n))$.

Since our algebras will be induced from K-theory we may assume that the orders Λ, Γ, Θ are closed under the Adams operations.

With all these assumptions on the orders Λ, Γ, Θ and $\alpha_\Gamma \in \Gamma$ we have reason to suspect that the answer to the following basic question is yes: Let $\varepsilon_\Lambda : \Lambda \longrightarrow \Lambda \underset{R}{\otimes} Q$ denote the obvious surjection. Here Q is made an R module by $a/b \longrightarrow a(1)/b(1) \in Q$.

<u>Basic Question</u>: <u>Is $\varepsilon_\Lambda f_*(1) = 1$?</u>

For the importance of this question, see II, Theorem 6.2 and the subsequence discussion.

Geometric Situation

We shall now apply the induction theory discussed above to
1 maps between closed S^1 manifolds.

If $[X,f] \in S_{S^1}(Y)$, we set

(i) $\Lambda = \tilde{K}^*_{S^1}(Y)$

(ii) $\Gamma = \tilde{K}^*_{S^1}(X)$

(iii) $\Theta = K^*_{S^1}(X^{S^1})_P = K^*_{S^1}(Y^{S^1})_P$

ecall that $\tilde{K}^*_{S^1}$ is localized equivariant K-theory mod $R = R(S^1)_P$
orsion. There is no R torsion in $K^*_{S^1}(X^{S^1})_P$ since
$^*_{S^1}(X^{S^1})_P = K^*(X^{S^1}) \otimes_Z R(S^1)_P$ and $R(S^1)_P$ contains the rationals.
lso, note that the second equality in (iii) follows from the
efinition of $S_{S^1}(Y)$: we required that $|f^{S^1}| : |X^{S^1}| \longrightarrow |Y^{S^1}|$
e a homotopy equivalence.

We have a commutative diagram of R algebra morphisms

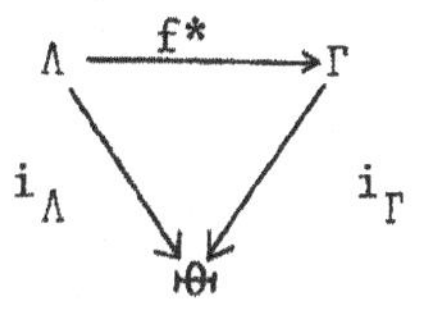

heorem 10.10: i_Λ, i_Γ, f^* are monomorphisms and induce
somorphism over $\bar{F}$. Moreover, f^*_p is an isomorphism for all
$\in P_1$.

Proof: This is a restatement of Proposition 5.1.

The S^1 tangent bundle of the smooth closed S^1 manifold X is denoted by TX. The Atiyah-Singer index homomorphism from $K^*_{S^1}(TX)$ to $R(S^1)$ induces an R homomorphism

$$Id^X: K^*_{S^1}(X)_P \longrightarrow K^*_{S^1}(TX)_P \quad [2]$$

and the composition $Id^X \circ \psi^X$ induces an R homomorphism Id^Γ

$$Id^\Gamma: \Gamma \longrightarrow R$$

Moreover, it follows from Part I, 6.19 that the R valued bilinear form on Γ defined by

$$\langle x,y\rangle_\Gamma = Id^\Gamma(x\cdot y) \qquad x,y \in \Gamma$$

is non-degenerate. This implies $(i_\Gamma)_*$ is defined. Set $\alpha_\Gamma = (i_\Gamma)_*$

It follows from 10.3 that the homomorphism Id^Γ has the form

$$Id^\Gamma(\gamma) = Id^\Theta\left(\frac{i_\Gamma(\gamma)}{i_\Gamma(\alpha_\Gamma)}\right), \quad \gamma \in \Gamma.$$

We want to relate α_Γ to the normal bundle of X^{S^1} in X.

Let N denote the S^1 normal bundle NX^{S^1} with a choice of a complex structure. Its underlying complex vector bundle $|N|$ has a spin^c structure. Since $|TX|$ has a spin^c structure it follows that $|TX^{S^1}|$ admits a spin^c structure given by the equation

$$|TX^{S^1}| \oplus |N| = |TX|\Big|_{|X^{S^1}|}$$

Therefore we have a Thom isomorphism

$$\psi^{X^{S^1}} : K^*_{S^1}(X^{S^1})_P \longrightarrow K^*_{S^1}(TX^{S^1})_P .$$

It follows from [5] that the index homomorphism Id^X is given by

$$Id^X(x) = Id^{X^{S^1}}((Ti)^*(x)/\lambda_{-1}(N\otimes C))$$

where $Ti: TX^{S^1} \longrightarrow TX$ is the inclusion. Moreover, by [2] we now have that

$$(\psi^{X^{S^1}})^{-1}(Ti)^*\psi^X(u) = i^*(u)\cdot\delta(N)$$

Here $\delta(N) \in \Theta$ is a class with the following property:

If N^* denotes the complex conjugate of N then $\delta(N)\cdot\delta(N^*) = \lambda_{-1}(N\otimes C) \in K^*_{S^1}(X^{S^1})_P = \Theta$ and $\delta(N^*) = \mu\cdot\lambda_{-1}(N) \in \Theta$ for some unit $\mu \in \Theta$.

In view of this information, we have for $\gamma \in \Gamma$

$$Id^\Gamma(\gamma) = Id^X\cdot\psi^X(\gamma) = Id^{X^{S^1}}\left(\frac{(Ti)^*\psi^X(\gamma)}{\lambda_{-1}(N\otimes C)}\right)$$

$$= Id^{X^{S^1}}\cdot\psi^{X^{S^1}}\cdot\{(\psi^{X^{S^1}})^{-1}(Ti)^*\psi^X\}\left(\frac{\gamma}{\lambda_{-1}(N\otimes C)}\right)$$

$$= \mathrm{Id}^{X^{S^1}} \cdot \psi^{X^{S^1}} \left(\frac{i^*(\gamma) \cdot \delta(N)}{\lambda_{-1}(N \otimes C)} \right)$$

$$= \mathrm{Id}^{X^{S^1}} \cdot \psi^{X^{S^1}} \left(\frac{i^*(\gamma)}{\delta(N^*)} \right) = \mathrm{Id}^{\Theta} \left(\frac{i_\Gamma(\gamma)}{\delta(N^*)} \right)$$

So by 10.3 $i_\Gamma(\alpha_\Gamma) = \delta(N^*) = \mu \cdot \lambda_{-1}(N)$, μ a unit in Θ.

Similarly $(i_\Lambda)_*(1) = \alpha_\Lambda$ is a unit in $\Lambda \otimes_R \bar{F}$ and $f_*(\alpha_\Gamma) = \alpha_\Lambda$. The results of induction can be applied now to $(i_\Lambda)_*$, $(i_\Gamma)_*$, f_* with $1 \in \Theta$, $\alpha_\Lambda \in \Lambda$, $\alpha_\Gamma \in \Gamma$ units in $\Lambda \otimes_R \bar{F} = \Gamma \otimes_R \bar{F} = \Theta \otimes_R \bar{F}$.

Denote by $\widehat{NY}^{S^1}$ a complex S^1 vector bundle whose underlying real S^1 vector bundle is NY^{S^1}. The relation between the image $f_*(1_X)$ of the unit element $1_X \in \Gamma$ and the normal bundles NX^{S^1} and NY^{S^1} is provided by Corollary 10.7 which gives

__Proposition 10.11:__ $i_\Lambda f_*(1_X) = \mu' \cdot \lambda_{-1}(\widehat{NY}^{S^1}) / \lambda_{-1}(\widehat{NX}^{S^1}) \in \Theta$ __where__ μ' __is a unit of__ Θ. __If__ $q \in Y^{S^1}$

$$i_\Lambda(\alpha_\Lambda)_q = \delta((\widehat{NY}^{S^1})^*_q) = \mu_q \cdot \lambda_{-1}(\widehat{NY}^{S^1}_q)$$

__for some unit__ μ_q __of__ R. __Here__ $i_\Lambda(\alpha_\Lambda)_q$ __denotes the restriction of__ $i_\Lambda(\alpha_\Lambda) \in \Theta$ __to__ q.

Here is a geometric consequence of the fact that the
ilinear form $< \; >_X$ is non-singular over R: Suppose that
$^{S^1}$ consists of isolated fixed points. For each $p \in X^{S^1}$ let

$$\widehat{TX}_p = \Sigma_{i=1}^{d/2} t^{\lambda_i(p)}$$

roposition 10.12: The absolute value of each integer $|\lambda_i(p)|$
ccurs an even number of times in the collection

$$\left\{ |\lambda_i(p)| \;\middle|\; \begin{matrix} p \in X^{S^1} \\ i = 1,2,\ldots d/2 \end{matrix} \right\}.$$

roof: Recall that $\mu \cdot \lambda_{-1}(\widehat{NX^{S^1}}) = \alpha_\Lambda = (i_\Lambda)_*(1)$, where $\mu \in \Lambda$
s a unit. By Proposition 10.5 $\det_\Lambda(\alpha_\Lambda) \in U(R) \cdot R^2$ so
$et_\Lambda(\lambda_{-1}(NX^{S^1})) \in U(R) \cdot R^2$. If $p_1, \ldots, p_k$ are the isolated
ixed points then

$$\det_\Lambda(\lambda_{-1}(NX^{S^1})) = \prod_i \prod_j (1 - t^{\lambda_i(p_j)}) \in U(R) \cdot R^2$$

o, up to sign, $\lambda_i(p)$ occurs an even number of times.

emark 10.15: Observe that 10.6, 10.7, and 10.11 relate the normal
undles NY^{S^1} and NX^{S^1} with the quotient Γ/Λ. This exhibits
he interesting interplay between algebra and geometry.

REFERENCES

[1] Atiyah, M.F., K-Theory, Benjamin, (1967).

[2] Atiyah, M.F. and Singer, I.M., The index of elliptic operators I and III, Ann. of Math. (2) 87 (1968) 484-530, 564-604.

[3] __________ and Bott, R., A Lefschetz fixed point formula for elliptic complexes II. Applications, Ann. of Math., (3) 88 (1968) 451-491.

[4] __________ and Hirzebruch, F., Spin manifolds and group actions, Essays on Topology and Related Topics, Ed. A. Haefliger and R. Narasinham, Springer-Verlag (1970).

[5] __________ and Segal, G., The index of elliptic operator II, Ann. of Math. (87) (1968) 531-545.

[6] ________________________, Equivariant K theory and completion, J. Diff. Geometry, (87) (1968) 531-545.

[7] Bredon, G., Introduction to Compact Transformation Groups Academic Press, (1972).

[8] __________, The cohomology ring structure of a fixed point set, Ann. of Math. (2) 80 (1964) 524-537.

[9] Milnor, J., Morse Theory, Ann. of Math. Study 51, Princet University Press (1963).

[10] __________, Singular Points of Complex Hypersurfaces, Ann. of Math. Studies 61, Princeton University Press (1968).

[11] Petrie, T., Smooth S^1 actions on homotopy complex projective spaces and related topics, Bull. AMS (2) 78 (1972) 105-153.

[12] __________, Exotic S^1 actions on CP^3 and related topics, Inventiones Math. 17 (1972) 317-327.

[13] __________, Torus actions on homotopy complex projective spaces, Inventiones Math., to appear.

[14] __________, Real algebraic actions on projective spaces a survey, Annales De L'INSTITUT FOURIER, 23, 2 (1973).

[15] __________, Smooth S^1 manifolds, to appear.

[16] __________, Induction in equivariant K theory and geomet applications, Seattle Conference on Algebraic K Theory, Springer-Verlag Lecture Series, to appear.

SYMBOL INDEX

Notation: G denotes throughout a compact Lie group and R(G) the complex representation ring of G. In particular, $R(S^1) = Z[t,t^{-1}]$ is the complex representation ring of the circle group S^1.

Part I

Part II

399: Functional Analysis and its Applications. Proceedings . Edited by H. G. Garnir, K. R. Unni and J. H. Williamson. 4 pages. 1974.

400: A Crash Course on Kleinian Groups. Proceedings 1974. d by L. Bers and I. Kra. VII, 130 pages. 1974.

401: M. F. Atiyah, Elliptic Operators and Compact Groups. pages. 1974.

402: M. Waldschmidt, Nombres Transcendants. VIII, 277 s. 1974.

403: Combinatorial Mathematics. Proceedings 1972. Edited A. Holton. VIII, 148 pages. 1974.

404: Théorie du Potentiel et Analyse Harmonique. Edité par aut. V, 245 pages. 1974.

405: K. J. Devlin and H. Johnsbråten, The Souslin Problem. 32 pages. 1974.

406: Graphs and Combinatorics. Proceedings 1973. Edited . A. Bari and F. Harary. VIII, 355 pages. 1974.

407: P. Berthelot, Cohomologie Cristalline des Schémas de cteristique p > o. II, 604 pages. 1974.

408: J. Wermer, Potential Theory. VIII, 146 pages. 1974.

409: Fonctions de Plusieurs Variables Complexes, Séminaire çois Norguet 1970–1973. XIII, 612 pages. 1974.

410: Séminaire Pierre Lelong (Analyse) Année 1972–1973. 31 pages. 1974.

411: Hypergraph Seminar. Ohio State University, 1972. d by C. Berge and D. Ray-Chaudhuri. IX, 287 pages. 1974.

412: Classification of Algebraic Varieties and Compact plex Manifolds. Proceedings 1974. Edited by H. Popp. V, pages. 1974.

413: M. Bruneau, Variation Totale d'une Fonction. XIV, 332 s. 1974.

414: T. Kambayashi, M. Miyanishi and M. Takeuchi, Unit Algebraic Groups. VI, 165 pages. 1974.

415: Ordinary and Partial Differential Equations. Proceedings XVII, 447 pages. 1974.

416: M. E. Taylor, Pseudo Differential Operators. IV, 155 s. 1974.

417: H. H. Keller, Differential Calculus in Locally Convex es. XVI, 131 pages. 1974.

418: Localization in Group Theory and Homotopy Theory Related Topics. Battelle Seattle 1974 Seminar. Edited by P. J. n. VI, 172 pages 1974.

419: Topics in Analysis. Proceedings 1970. Edited by O. E. , I. S. Louhivaara, and R. H. Nevanlinna. XIII, 392 pages. 1974.

420: Category Seminar. Proceedings 1972/73. Edited by G. M. VI, 375 pages. 1974.

421: V. Poénaru, Groupes Discrets. VI, 216 pages. 1974.

422: J.-M. Lemaire, Algèbres Connexes et Homologie des ces de Lacets. XIV, 133 pages. 1974.

423: S. S. Abhyankar and A. M. Sathaye, Geometric Theory gebraic Space Curves. XIV, 302 pages. 1974.

424: L. Weiss and J. Wolfowitz, Maximum Probability ators and Related Topics. V, 106 pages. 1974.

425: P. R. Chernoff and J. E. Marsden, Properties of Infinite nsional Hamiltonian Systems. IV, 160 pages. 1974.

426: M. L. Silverstein, Symmetric Markov Processes. X, 287 s. 1974.

427: H. Omori, Infinite Dimensional Lie Transformation ps. XII, 149 pages. 1974.

428: Algebraic and Geometrical Methods in Topology, Prongs 1973. Edited by L. F. McAuley. XI, 280 pages. 1974.

Vol. 429: L. Cohn, Analytic Theory of the Harish-Chandra C-Function. III, 154 pages. 1974.

Vol. 430: Constructive and Computational Methods for Differential and Integral Equations. Proceedings 1974. Edited by D. L. Colton and R. P. Gilbert. VII, 476 pages. 1974.

Vol. 431: Séminaire Bourbaki – vol. 1973/74. Exposés 436–452. IV, 347 pages. 1975.

Vol. 432: R. P. Pflug, Holomorphiegebiete, pseudokonvexe Gebiete und das Levi-Problem. VI, 210 Seiten. 1975.

Vol. 433: W. G. Faris, Self-Adjoint Operators. VII, 115 pages. 1975.

Vol. 434: P. Brenner, V. Thomée, and L. B. Wahlbin, Besov Spaces and Applications to Difference Methods for Initial Value Problems. II, 154 pages. 1975.

Vol. 435: C. F. Dunkl and D. E. Ramirez, Representations of Commutative Semitopological Semigroups. VI, 181 pages. 1975.

Vol. 436: L. Auslander and R. Tolimieri, Abelian Harmonic Analysis, Theta Functions and Function Algebras on a Nilmanifold. V, 99 pages. 1975.

Vol. 437: D. W. Masser, Elliptic Functions and Transcendence. XIV, 143 pages. 1975.

Vol. 438: Geometric Topology. Proceedings 1974. Edited by L. C. Glaser and T. B. Rushing. X, 459 pages. 1975.

Vol. 439: K. Ueno, Classification Theory of Algebraic Varieties and Compact Complex Spaces. XIX, 278 pages. 1975

Vol. 440: R. K. Getoor, Markov Processes: Ray Processes and Right Processes. V, 118 pages. 1975.

Vol. 441: N. Jacobson, PI-Algebras. An Introduction. V, 115 pages. 1975.

Vol. 442: C. H. Wilcox, Scattering Theory for the d'Alembert Equation in Exterior Domains. III, 184 pages. 1975.

Vol. 443: M. Lazard, Commutative Formal Groups. II, 236 pages. 1975.

Vol. 444: F. van Oystaeyen, Prime Spectra in Non-Commutative Algebra. V, 128 pages. 1975.

Vol. 445: Model Theory and Topoi. Edited by F. W. Lawvere, C. Maurer, and G. C. Wraith. III, 354 pages. 1975.

Vol. 446: Partial Differential Equations and Related Topics. Proceedings 1974. Edited by J. A. Goldstein. IV, 389 pages. 1975.

Vol. 447: S. Toledo, Tableau Systems for First Order Number Theory and Certain Higher Order Theories. III, 339 pages. 1975.

Vol. 448: Spectral Theory and Differential Equations. Proceedings 1974. Edited by W. N. Everitt. XII, 321 pages. 1975.

Vol. 449: Hyperfunctions and Theoretical Physics. Proceedings 1973. Edited by F. Pham. IV, 218 pages. 1975.

Vol. 450: Algebra and Logic. Proceedings 1974. Edited by J. N. Crossley. VIII, 307 pages. 1975.

Vol. 451: Probabilistic Methods in Differential Equations. Proceedings 1974. Edited by M. A. Pinsky. VII, 162 pages. 1975.

Vol. 452: Combinatorial Mathematics III. Proceedings 1974. Edited by Anne Penfold Street and W. D. Wallis. IX, 233 pages. 1975.

Vol. 453: Logic Colloquium. Symposium on Logic Held at Boston, 1972–73. Edited by R. Parikh. IV, 251 pages. 1975.

Vol. 454: J. Hirschfeld and W. H. Wheeler, Forcing, Arithmetic, Division Rings. VII, 266 pages. 1975.

Vol. 455: H. Kraft, Kommutative algebraische Gruppen und Ringe. III, 163 Seiten. 1975.

Vol. 456: R. M. Fossum, P. A. Griffith, and I. Reiten, Trivial Extensions of Abelian Categories. Homological Algebra of Trivial Extensions of Abelian Categories with Applications to Ring Theory. XI, 122 pages. 1975.

Vol. 457: Fractional Calculus and Its Applications. Proceedings 1974. Edited by B. Ross. VI, 381 pages. 1975.

Vol. 458: P. Walters, Ergodic Theory – Introductory Lectures. VI, 198 pages. 1975.

Vol. 459: Fourier Integral Operators and Partial Differential Equations. Proceedings 1974. Edited by J. Chazarain. VI, 372 pages. 1975.

Vol. 460: O. Loos, Jordan Pairs. XVI, 218 pages. 1975.

Vol. 461: Computational Mechanics. Proceedings 1974. Edited by J. T. Oden. VII, 328 pages. 1975.

Vol. 462: P. Gérardin, Construction de Séries Discrètes p-adiques. »Sur les séries discrètes non ramifiées des groupes réductifs déployés p-adiques«. III, 180 pages. 1975.

Vol. 463: H.-H. Kuo, Gaussian Measures in Banach Spaces. VI, 224 pages. 1975.

Vol. 464: C. Rockland, Hypoellipticity and Eigenvalue Asymptotics. III, 171 pages. 1975.

Vol. 465: Séminaire de Probabilités IX. Proceedings 1973/74. Edité par P. A. Meyer. IV, 589 pages. 1975.

Vol. 466: Non-Commutative Harmonic Analysis. Proceedings 1974. Edited by J. Carmona, J. Dixmier and M. Vergne. VI, 231 pages. 1975.

Vol. 467: M. R. Essén, The Cos $\pi\lambda$ Theorem. With a paper by Christer Borell. VII, 112 pages. 1975.

Vol. 468: Dynamical Systems – Warwick 1974. Proceedings 1973/74. Edited by A. Manning. X, 405 pages. 1975.

Vol. 469: E. Binz, Continuous Convergence on C(X). IX, 140 pages. 1975.

Vol. 470: R. Bowen, Equilibrium States and the Ergodic Theory of Anosov Diffeomorphisms. III, 108 pages. 1975.

Vol. 471: R. S. Hamilton, Harmonic Maps of Manifolds with Boundary. III, 168 pages. 1975.

Vol. 472: Probability-Winter School. Proceedings 1975. Edited by Z. Ciesielski, K. Urbanik, and W. A. Woyczyński. VI, 283 pages. 1975.

Vol. 473: D. Burghelea, R. Lashof, and M. Rothenberg, Groups of Automorphisms of Manifolds. (with an appendix by E. Pedersen) VII, 156 pages. 1975.

Vol. 474: Séminaire Pierre Lelong (Analyse) Année 1973/74. Edité par P. Lelong. VI, 182 pages. 1975.

Vol. 475: Répartition Modulo 1. Actes du Colloque de Marseille-Luminy, 4 au 7 Juin 1974. Edité par G. Rauzy. V, 258 pages. 1975. 1975.

Vol. 476: Modular Functions of One Variable IV. Proceedings 1972. Edited by B. J. Birch and W. Kuyk. V, 151 pages. 1975.

Vol. 477: Optimization and Optimal Control. Proceedings 1974. Edited by R. Bulirsch, W. Oettli, and J. Stoer. VII, 294 pages. 1975.

Vol. 478: G. Schober, Univalent Functions – Selected Topics. V, 200 pages. 1975.

Vol. 479: S. D. Fisher and J. W. Jerome, Minimum Norm Extremals in Function Spaces. With Applications to Classical and Modern Analysis. VIII, 209 pages. 1975.

Vol. 480: X. M. Fernique, J. P. Conze et J. Gani, Ecole d'Eté de Probabilités de Saint-Flour IV–1974. Edité par P.-L. Hennequin. XI, 293 pages. 1975.

Vol. 481: M. de Guzmán, Differentiation of Integrals in R^n. XII, 226 pages. 1975.

Vol. 482: Fonctions de Plusieurs Variables Complexes II. Séminaire François Norguet 1974–1975. IX, 367 pages. 1975.

Vol. 483: R. D. M. Accola, Riemann Surfaces, Theta Functions, and Abelian Automorphisms Groups. III, 105 pages. 1975.

Vol. 484: Differential Topology and Geometry. Proceedings 1974. Edited by G. P. Joubert, R. P. Moussu, and R. H. Roussarie. IX, 287 pages. 1975.

Vol. 485: J. Diestel, Geometry of Banach Spaces – Selected To XI, 282 pages. 1975.

Vol. 486: S. Stratila and D. Voiculescu, Representations of Algebras and of the Group U(∞). IX, 169 pages. 1975.

Vol. 487: H. M. Reimann und T. Rychener, Funktionen beschrä mittlerer Oszillation. VI, 141 Seiten. 1975.

Vol. 488: Representations of Algebras, Ottawa 1974. Procee 1974. Edited by V. Dlab and P. Gabriel. XII, 378 pages. 1975.

Vol. 489: J. Bair and R. Fourneau, Etude Géométrique des Esp Vectoriels. Une Introduction. VII, 185 pages. 1975.

Vol. 490: The Geometry of Metric and Linear Spaces. Procee 1974. Edited by L. M. Kelly. X, 244 pages. 1975.

Vol. 491: K. A. Broughan, Invariants for Real-Generated Un Topological and Algebraic Categories. X, 197 pages. 1975.

Vol. 492: Infinitary Logic: In Memoriam Carol Karp. Edited by Kueker. VI, 206 pages. 1975.

Vol. 493: F. W. Kamber and P. Tondeur, Foliated Bundles Characteristic Classes. XIII, 208 pages. 1975.

Vol. 494: A Cornea and G. Licea. Order and Potential Resc Families of Kernels. IV, 154 pages. 1975.

Vol. 495: A. Kerber, Representations of Permutation Groups 175 pages. 1975.

Vol. 496: L. H. Hodgkin and V. P. Snaith, Topics in K-Theory Independent Contributions. III, 294 pages. 1975.

Vol. 497: Analyse Harmonique sur les Groupes de Lie. Procee 1973–75. Edité par P. Eymard et al. VI, 710 pages. 1975.

Vol. 498: Model Theory and Algebra. A Memorial Tribu Abraham Robinson. Edited by D. H. Saracino and V. B. Weispfer X, 463 pages. 1975.

Vol. 499: Logic Conference, Kiel 1974. Proceedings. Edite G. H. Müller, A. Oberschelp, and K. Potthoff. V, 651 pages

Vol. 500: Proof Theory Symposion, Kiel 1974. Proceedings. Edi J. Diller and G. H. Müller. VIII, 383 pages. 1975.

Vol. 501: Spline Functions, Karlsruhe 1975. Proceedings. Edit K. Böhmer, G. Meinardus, and W. Schempp. VI, 421 pages. 1976.

Vol. 502: János Galambos, Representations of Real Numbe Infinite Series. VI, 146 pages. 1976.

Vol. 503: Applications of Methods of Functional Analysis to Prob in Mechanics. Proceedings 1975. Edited by P. Germain ar Nayroles. XIX, 531 pages. 1976.

Vol. 504: S. Lang and H. F. Trotter, Frobenius Distributior GL_2-Extensions. III, 274 pages. 1976.

Vol. 505: Advances in Complex Function Theory. Procee 1973/74. Edited by W. E. Kirwan and L. Zalcman. VIII, 203 p 1976.

Vol. 506: Numerical Analysis, Dundee 1975. Proceedings. E by G. A. Watson. X, 201 pages. 1976.

Vol. 507: M. C. Reed, Abstract Non-Linear Wave Equations 128 pages. 1976.

Vol. 508: E. Seneta, Regularly Varying Functions. V, 112 pages. 1

Vol. 509: D. E. Blair, Contact Manifolds in Riemannian Geon VI, 146 pages. 1976.

Vol. 510: V. Poènaru, Singularités C^∞ en Présence de Sym V, 174 pages. 1976.

Vol. 511: Séminaire de Probabilités X. Proceedings 1974/75. par P. A. Meyer. VI, 593 pages. 1976.

Vol. 512: Spaces of Analytic Functions, Kristiansand, Norway Proceedings. Edited by O. B. Bekken, B. K. Øksendal, and A. VIII, 204 pages. 1976.

Vol. 513: R. B. Warfield, Jr. Nilpotent Groups. VIII, 115 pages.

Vol. 514: Séminaire Bourbaki vol. 1974/75. Exposés 453 – 4 276 pages. 1976.

Vol. 515: Bäcklund Transformations. Nashville, Tennessee Proceedings. Edited by R. M. Miura. VIII, 295 pages. 1976.